Letting God Create Your Day

Gen. 1:1

Steve Schwartz

Volume 8
Steve Schwartz

The Scripture quotations in this publication are from **The Holy Bible, King James Version,** World Publishing Company.

**Letting God Create Your Day: Volume 8
by Steve Schwartz**

Copyright © 2016 Creation Moments, Inc.

Creation Moments, Inc.
P.O. Box 839 Foley, Minnesota 56329
www.creationmoments.com
800-422-4253

ISBN 1-882510-22-4

All rights reserved. No portion of this book may be reproduced, stored in a retrieval system, or transmitted in any form or by any means, electronic, mechanical, photocopying, recording, or otherwise, without prior written permission from the publisher.

Printed in the United States of America
Printing and production costs for this book were underwritten by supporters of Creation Moments.

Cover: Image used under license from Shutterstock.com

Foreword

Two evolutionists and a creationist are walking through the park on a beautiful sunny day. As could be expected, they are at an impasse in their heated discussion about evolution and biblical creation. Finally, out of frustration, the creationist asks the evolutionists if it's okay with them if he prays for a resolution to their stalemate. They laugh at him, saying they don't believe in God anyway, but they tell him to go ahead.

Moments after the creationist prays, storm clouds gather overhead, lightning flashes across the darkened sky, and a thunderous voice comes out of the clouds: "The creationist is right!"

The creationist, of course, is overjoyed ... until one of the evolutionists says, "So? It's still only a tie vote."

Why am I telling you this story? For one thing, I am using it to make the point that evolutionists cannot be convinced against their will that the Bible's account of creation is true. As the saying goes: "A man convinced against his will ... is of the same opinion still." Only the Holy Spirit can change someone's heart and make him willing to approach the Bible in complete confidence that it really is the inspired Word of God and authoritative source of truth.

The second reason for the story is to point out the foolishness of thinking that God's thoughts have equal weight to men's thoughts. As God tells us in the Bible, "For my thoughts are not your thoughts, neither are your ways my ways, saith the LORD. For as the heavens are higher than the earth, so are my ways higher than your ways, and my thoughts than your thoughts" (Isaiah 55:8-9).

So if the Bible truly is the inspired Word of God, the creation account found in the book of Genesis has infinitely more weight than the opinions of a million scientists. If the Creator Himself told us how He created the heavens and the earth, how foolish it is to ignore His eyewitness testimony.

With these things in mind, I think I can safely say you are going to love this volume of *Letting God Create Your Day* if you believe that the Bible can be trusted.

On the other hand, perhaps you aren't quite sure whether to trust what the Bible says about creation. Or you might even be wondering if evolution and the Bible's account of creation could *both* be true. Let me just say that evolution and biblical creation are *totally* incompatible. And that's not just my opinion. This is one point on which most evolutionists and creationists agree.

If you are unsure of what to believe, here are a few suggestions. As you read this book, don't skip the Bible verses you'll find at the top of each page. God can reveal important truths to you as you read them. Also, use the prayer at the bottom of each page to launch a prayer from your own heart. Ask God to remove your confusion and make the truth crystal clear.

In addition, I urge you to do more research on any of the topics covered in this book. When a topic is particularly interesting to you, look up the references at the bottom of each page. You'll also find a great deal of useful information at CreationMoments.com.

I close by saying that for the past two years, it has been both an honor and a pleasure to write the scripts for the Creation Moments radio program. I am grateful to Ian Taylor for speaking my words so wonderfully, and I am thankful to Creation Moments for giving me the opportunity to write the scripts you're about to read. I hope that these scripts are as much a blessing to you as they were to me in writing them.

To God be the glory!

Steve Schwartz
October, 2016

Living Gears!

Psalm 5:8
"Lead me, O LORD, in thy righteousness because of mine enemies; make thy way straight before my face."

Mechanical gears – like those found in clocks – have been around since the Greeks are thought to have invented them around 300 BC. But scientists have now discovered a small hopping insect equipped with a set of *living* gears!

University of Cambridge biologists discovered that *Issus coleoptratus* have an intricate gearing system that locks their back legs together. This allows both legs to spring at the exact same instant, propelling the tiny creatures straight forward. If one of the bug's legs jumped a fraction of a second earlier than the other, this would push the insect off course to the left or right.

The gears are located at the top of the insects' hind legs and include ten to twelve tapered teeth. The teeth of the gears lock together neatly, and they even have curves at the base, a design incorporated into man-made mechanical gears to reduce wear over time.

Researcher Gregory Sutton said, "We usually think of gears as something that we see in human-designed machinery.... These gears are not designed; they are evolved – representing high speed and precision machinery evolved for synchronization in the animal world."

What we'd like to ask him is how did this insect survive for thousands of years while it couldn't jump straight? No, there is a much simpler explanation that scientists might see if they weren't so biased against a Creator. The gears were designed by God, who gave all of His creatures – including you and me – all of the intricate parts we need.

Prayer: Lord Jesus, I praise You and thank You for making it so easy to see Your handprints on the things You have made. I pray that many scientists will come to realize that there is a Designer ... and that Designer is You! Amen.

Ref: Joseph Stromberg, "This Insect Has The Only Mechanical Gears Ever Found in Nature," Smithsonian.com, 9/12/13.

The Chicken from Hell

1 Corinthians 15:39
"All flesh is not the same flesh: but there is one kind of flesh of men, another flesh of beasts, another of fishes, and another of birds."

Evolutionists tell us a 66-million-year-old feathered dinosaur resembling a giant demonic bird was discovered in the fossil-rich Hell Creek formation of South and North Dakota. Not surprisingly, this newly discovered dinosaur species was dubbed the "chicken from hell."

It was "as close as you can get to a bird without being a bird," said vertebrate paleontologist Matt Lamanna. In addition to its long limbs, the research team found that the 11-foot-long, 500-pound dinosaur sported a stubby tail "likely framed by a fan of tail feathers."

What's this? A tail *likely framed by a fan of tail feathers?* Not even the writers at *National Geographic* could stop themselves from telling the truth when they admitted: "Though the team *didn't find direct evidence of feathers*, the species was so closely related to birds that it was *very likely* covered in feathers that looked identical to those of modern birds."

The illustration accompanying the article shows the dinosaur's forelimbs and tail covered with large feathers. Indeed, *National Geographic* and others have been doing this for years as they attempt to show that modern-day birds are the evolutionary descendents of dinosaurs. No real evidence required!

The Bible, on the other hand, is supported by a great amount of evidence, including archaeological discoveries, the fulfillment of Bible prophecies and much more. Christians also have the internal witness of the Holy Spirit that convinces us that God's Word is true and that His Son is the Truth who alone provides salvation.

Prayer: Father, I thank You that Your Word doesn't need to be supported by the imaginative speculations of fallible men. In Jesus' Name. Amen.

Ref: Christine Dell'Amore, "New 'Chicken From Hell' Dinosaur Discovered," *National Geographic Daily News*, 3/19/14.

When Being "Wrong" Is Right

Proverbs 1:7
"The fear of the LORD is the beginning of knowledge: but fools despise wisdom and instruction."

How smart are Americans when it comes to science? That's what the National Science Foundation attempts to find out every two years. The results of their 2014 survey – which included more than 2,200 adults – gave Americans a rather poor grade. But are Americans really less knowledgeable about science or is the NSF's survey biased against the many Bible-believing Christians in America? Let's take a look.

While the survey included such questions as "Does the Earth go around the Sun or does the Sun go around the Earth?", the survey also included questions like: "True or false: the universe began with a huge explosion." And true or false: "Human beings, as we know them today, developed from earlier species of animals." For people who believe what the Bible has to say, their answers were scored as "wrong" by the NSF survey.

Interestingly, more than 60 percent of Americans gave the so-called "wrong" answers to these questions. They simply do not accept the Big Bang theory despite what they were taught in school. And more than 50 percent of Americans disagreed with the NSF when it comes to evolution. According to the *Atlantic Online*, "This seems to indicate that many Americans are familiar with the theories of evolution and the Big Bang; they simply don't believe they're true."

What the unbelieving world views as right is often wrong in light of what God tells us in the Bible. For Christians, the Bible is the ultimate authority – not what we're told by Bible-denying evolutionists.

Prayer: Heavenly Father, I pray that You will strengthen me whenever I am pressured by the world to choose wrong over right. Make me unafraid to be thought a fool in the world's eyes. In Jesus' Name. Amen.

Ref: Eleanor Barkhorn, "What Americans Don't Know About Science," *Atlantic* online. 2/15/14.

Plants Really Do Talk to Each Other

Judges 9:10
"And the trees said to the fig tree, Come thou, and reign over us."

For many years, Creation Moments has been sharing information about the ways that plants and trees "talk" to one another so they can protect each other from invaders. Many scientists have ridiculed us for making such "outrageous" claims. But they're not laughing any longer.

As *Wired* online recently commented, "The evidence for plant communication is only a few decades old. The first few 'talking tree' papers quickly were shot down as statistically flawed or too artificial and research ground to a halt. But the science of plant communication is now staging a comeback."

Though plant communication is still a tiny field, the scientists who study it are no longer seen as a lunatic fringe. Richard Karban, an ecologist at the University of California in Davis, said: "It used to be that people wouldn't even talk to you: 'Why are you wasting my time with something we've already debunked?' they said."

But scientists are now saying that plants not only communicate with other plants, they communicate with insects as well. They send airborne messages that act as distress signals to predatory insects so that they will come and kill the plant-eating bugs!

Though scientists are now learning more about how plants communicate, many appear to have little interest in learning about how God communicates with man. But scientists who are Christians know the blessing of communicating with God through prayer and the reading of His inspired Word.

Prayer: Father, I am so grateful that You care about us and that You communicate with us through the Bible. Remind me to make my prayer time and the study of Your Word a top priority in my life. In Jesus' Name. Amen.

Ref: "How Plants Secretly Talk to Each Other," *Wired* online, 12/20/13.

Bats Cry: That's Mine!

Ephesians 4:28
"Let him that stole steal no more: but rather let him labour, working with his hands the thing which is good, that he may have to give to him that needeth."

As you know, bats are able to find their way around and locate their prey using echolocation – high-pitched sounds that they also use to keep from flying into trees and other bats. But scientists have now learned that one particular kind of bat emits another call that warns other bats to stay away from bugs they've claimed for themselves.

A research team from the University of Maryland recently discovered that male big brown bats produce a special sound called a frequency-modulated bout, or FMB, that sounds a warning to other bats. The FMB is an ultrasonic social call that uniquely identifies the bat emitting it. Its sequence of three to four sounds is longer in duration and lower in frequency than typical echolocation pulses that the bats use for navigation. After hearing the FMB, other bats moved away from both the caller and its lunch.

Biology Research Associate Genevieve Spanjer Wright said: "When two males flew together in a trial, it was not uncommon for each bat to emit FMBs. We found that the bat emitting the greatest number of FMBs was more likely to capture the mealworm."

While only a few talented humans have learned to use echolocation, all of us have been blessed by our Creator with a conscience that acts like a warning signal from God. Let us be like the male big brown bat that not only heeds these warnings but also warns others to avoid things that would displease our Creator.

Prayer: Heavenly Father, I pray that You will make my conscience shout at me when I am about to do anything that displeases You. In Jesus' Name. Amen.

Ref: "Foraging bats can warn each other away from their dinners," 3/27/14. *ScienceDaily* online. Genevieve Spanjer Wright, Chen Chiu, Wei Xian, Gerald S. Wilkinson, Cynthia F. Moss. Social Calls Predict Foraging Success in Big Brown Bats. *Current Biology*, 2014; DOI: 10.1016/j.cub.2014.02.058.

New Type of Eye Discovered!

Psalm 119:18
"Open thou mine eyes, that I may behold wondrous things out of thy law."

The University of Tübingen's Institute of Anatomy recently discovered a fish with what they called "a previously unknown type of eye." The glasshead barreleye has a cylindrical eye pointing upwards to see prey, predators and potential mates. But the eye also has a mirror-like second retina which is able to detect bioluminescent flashes created by deep-sea denizens to the sides and below, reports Professor Hans-Joachim Wagner.

Actually, it's not really a previously unknown type of eye. Reflector eyes are usually only found in invertebrates, such as mollusks and crustaceans, but there is one other vertebrate that also uses a combination of reflective and refractive lenses in its eyes – the deep-sea brownsnout spookfish.

Now this is where it gets really interesting. According to the report, both the glasshead barreleye and the brownsnout spookfish developed this amazingly complex eye *from different kinds of tissue.* So if you're going to believe in evolution, you'll have to deal with the fact that this unusual type of eye had to evolve twice, independently of one another, using two different solutions to the same problem.

Those of us who accept biblical creation often wonder at the blindness and stubbornness of evolutionary scientists. How can they continue to cling to random mutations and natural selection when only a supremely intelligent Creator could account for the incredibly complex eye of the glasshead barreleye?

Prayer: Father, I thank You for my eyes that enable me to see the world around me. Thank You also for giving me eyes of faith to see the work Your Son has done to secure eternal life for me. Amen.

Ref: "Researchers discover fish with a previously unknown type of eye," *Phys.Org,* 3/20/14.

Did Comets Help Create Life?

Psalm 147:16-17
"He giveth snow like wool: he scattereth the hoarfrost like ashes. He casteth forth his ice like morsels: who can stand before his cold?"

Would you believe that life on planet Earth was the result of icy comets striking the Earth's surface long, long ago? No, we didn't think so. But a team of researchers from three distinguished institutions would disagree with you.

To understand the effect of a comet hitting a planet, the researchers fired ice projectiles through a high-speed gun at targets that they said had compounds similar to those found on comets. When the ice chips stopped flying, they discovered that the impacts had created some amino acids.

One researcher wrote that "this process demonstrates a very simple mechanism whereby we can go from a mix of simple molecules, such as water and carbon-dioxide ice, to a more complicated molecule, such as an amino acid. This," he wrote, "is the first step towards life. The next step is to work out how to go from an amino acid to even more complex molecules such as proteins."

Good luck with that! Scientists have been attempting to do that since 1953, when Stanley Miller and Harold Urey at the University of Chicago produced amino acids by shooting bolts of simulated lightning into a concoction of chemicals. Proteins – the real building blocks of life – are light-years away from amino acids.

How sad that scientists are wasting so much time looking for answers to how life began. If only they believed the Bible's account of origins, they could dedicate their skills to pursuing real scientific achievements that benefit mankind.

Prayer: Heavenly Father, I ask You to lead more Christians into the sciences so they can do research that honors You and benefits others. In Jesus' Name. Amen.

Ref: "Comets Hitting Planets May Help Build Life, Study Finds," *Nature World News*, 9/16/13.

Darwin's Dilemma Solved?

Genesis 2:1
"Thus the heavens and the earth were finished, and all the host of them."

Darwin's dilemma has been vexing evolutionists for more than 150 years. Charles Darwin, you see, was unable to explain the sudden arrival in Cambrian rock of most of the phyla we have today when no fossil evidence of their ancestors can be found in pre-Cambrian rock.

Not surprisingly, creationists frequently mention this *sudden* appearance of complex life forms when challenging the Darwinian view of the *gradual* origin of species. So evolutionists have been struggling to find a plausible answer.

Such an answer was recently proposed by researchers at South Australia's University of Adelaide. They tried to get around the dilemma by estimating that the rate of evolution during the Cambrian explosion was five times faster than it is today.

The evolutionary website *ScienceDaily* states with great confidence, "The findings ... resolve 'Darwin's dilemma': the sudden appearance of a plethora of modern animal groups in the fossil record during the early Cambrian period."

But is the assertion that evolution happened five times faster than today supported by any evidence? No, it isn't. In fact, it only creates another dilemma for evolutionists who have always insisted on *uniformitarianism* – that the rates of natural processes do not change over time. Darwin's dilemma remains unsolved for evolutionists. But there is no dilemma at all for creationists who know that God created all living creatures on days five and six of Creation Week.

Prayer: Lord Jesus, I ask You to open the eyes of evolutionary scientists who reject what the Bible so clearly teaches. Amen.

Ref: Michael S.Y. Lee, Julien Soubrier, Gregory D. Edgecombe, "Rates of Phenotypic and Genomic Evolution during the Cambrian Explosion," *Current Biology*, 2013 DOI: 10.1016/j.cub.2013.07.055. "Darwin's dilemma resolved: Evolution's 'big bang' explained by five times faster rates of evolution," *ScienceDaily*, 9/12/13.

Fascinatin' Rhythm

Deuteronomy 6:5
"And thou shalt love the LORD thy God with all thine heart, and with all thy soul, and with all thy might."

The heart is an amazingly complex organ consisting of muscle tissue. And what do muscles do? They contract when stimulated by electric pulses. Though the human heart will beat 2-3 billion times in a person's lifetime, heart disease is one of the most common causes of death. That's why researchers have been looking for a way to produce artificial heart tissue. Have they finally succeeded?

Lei Yang and his team of researchers at the University of Pittsburgh reported that they were able to transform human skin into lab-grown "heart muscle cells" that contracted spontaneously in a dish.

To accomplish this extraordinary feat, the researchers first changed skin cells into stem cells and then treated them with growth factors. Then they added the cells to the structural shell of a mouse heart that had been stripped of its own cells. The human stem cells transformed the mouse heart shell into tissue that pulsed at 40 to 50 beats per minute.

Evolutionists, no doubt, will look at this achievement as making headway in man's quest to create artificial life. But let us remind them that the cells were produced using already-existing skin cells, mouse cells and growth factors with "design" built in. Besides, the "heart cells" were not a product of purposeless chance or mutations but the result of considerable effort by an *intelligent* team of scientists ... a reminder that life itself is the work of an intelligent, purposeful and all-wise Creator.

Prayer: Father, I thank You for my heart that sustains my life by beating regularly for so many years. I praise You most of all for turning my heart of stone into a heart that loves You, both now and long after my physical heart has stopped beating. In Jesus' Name. Amen.

Ref: *ScienceNews*, Web Edition 8/15/13. J. Shugart, "Lab-grown heart has rhythm." *Nature Communications 4*, Article #2307. "Repopulation of decellularized mouse heart with human induced pluripotent stem cell-derived cardiovascular progenitor cells," 8/13/13.

The Strongest Material in the World

Job 40:18
"His bones are as strong pieces of brass; his bones are like bars of iron."

Graphene may be the most unusual material that will ever be produced. It has already revolutionized the world of physics and won the 2010 Nobel Prize in physics for physicist Andre Geim and his colleague, Kostya Novoselow.

What's so unusual about graphene? Geim told *CNN Tech*, "It's the thinnest material you can get – it's only one atom thick. It's the strongest material we are aware of because you can't slice it any further. You can't get any material that is thinner than one atom, or it wouldn't count as a material anymore."

Geim went on to say, "Graphene is stronger than diamond; it shows extraordinary heat conductance; it conducts electricity a thousand times better than copper – the list goes on. Another surprise is that you can just about see it with the naked eye, even though it's only one atom thick!"

Geim added, "We live in a three-dimensional world. My physics intuition, developed over the last thirty years, told me that this material shouldn't exist. And if you had asked 99.9% of scientists around the world, they would have said the idea of a 2D material was rubbish and that graphene shouldn't exist."

How true that is! Scientists have been mistaken about a number of things, even when 99.9% of them were in agreement. Vestigial organs in humans, for example. Cosmic ether, for another. Then there's the fire-like element called phlogiston and the mistaken notion of epicycles. But the biggest mistake that scientists continue to make is evolution.

Prayer: Heavenly Father, let me never compromise with the world when I'm told that Your Word is untrue. If I am to be ridiculed or persecuted for following You, make me willing to be thought a fool for Christ's sake. Amen.

Ref: Stefanie Blendis, "Graphene: 'Miracle material' will be in your home sooner than you think," CNN, 10/6/13.

"Creation Is a Scientific Fact"

Isaiah 53:7
"He was oppressed, and he was afflicted, yet he opened not his mouth: he is brought as a lamb to the slaughter, and as a sheep before her shearers is dumb, so he openeth not his mouth."

In March of 2014 a team of scientists headed by astronomer John M. Kovac of the Harvard-Smithsonian Center for Astrophysics announced that they had finally found the evidence they were seeking to support the theory that the universe had a beginning.

This news about the Big Bang caused a Jewish physics professor, Nathan Aviezer of Bar Ilan University, to say: "One thing the announcement does do is make it clear that the universe had a definite starting point – a creation – as described in the book of Genesis."

He went on to say, "The Torah quotes God as saying 'let there be light,' and science tells us that this light came into existence, exploding to create the universe as we know it. ... At this point I think we can say that creation is a scientific fact. Without addressing who or what caused it, the mechanics of the creation process in the Big Bang match the Genesis story perfectly," he said. "If I had to make up a theory to match the first passages in Genesis, the Big Bang theory would be it."

Let us not forget, however, that the Big Bang theory is still just that – a theory – and there are many scientists who reject it. In fact, we will talk more about Big-Bang-denying scientists in a future broadcast. But for now we'll simply say that, Big Bang or not, we rejoice that some scientists are coming to realize that the concept of a Creator is not anti-science.

Prayer: Heavenly Father, I pray that more Jewish people would believe not only in biblical creation but in the Messiah, whose substitutionary death was foretold so clearly in the 53rd chapter of Isaiah. Amen.

Ref: "Physicist: Big Bang Breakthrough Confirms 'Creation'," WND.com, 3/19/14.

Evolutionist Rebukes Evolutionists

1 Thessalonians 5:21
"Prove all things; hold fast that which is good."

Meet Wolf-Ekkehard Lönnig, a brilliant scientist who worked at the Max-Planck Institute until his retirement. Though he has written four books on the subject of evolution, he has infuriated evolutionists everywhere by daring to challenge Neo-Darwinism on scientific grounds.

He told the *Diplomacy Post* in March 2014: "A scientific hypothesis should be potentially falsifiable.... However," he added, "the idea of slow evolution by 'infinitesimally small inherited variations' etc. has been falsified by the findings of palaeontology... as well [as] genetics. Yet its adherents principally reject any scientific proof against Neo-Darwinism, so that, in fact, their theory has become a non-falsifiable worldview, to which people stick in spite of all contrary evidence."

Scientists continue to support evolution despite the evidence that actually falsified evolution because "without Darwinism, philosophic materialism has lost its battle against an intelligent origin of the world."

But Dr. Lönnig had more to say. "According to Neo-Darwinism, all important problems of the origin of species are, at least in principle, solved. Further questions on the validity of evolutionary theory are thus basically superfluous. However, such a dogmatic attitude stops further investigations and hinders fruitful research in science."

Though Wolf-Ekkehard Lönnig has stopped short of giving recognition to God as the Intelligent Designer, we applaud his work for showing that evolution doesn't even deserve to be called scientific.

Prayer: Father, I pray for Intelligent Design proponents. Though they understand that the creation requires an intelligent designer, many of them still don't realize that the Intelligent Designer has a name – Jesus! Amen.

Ref: "Lönnig: Complex systems in nature point to an intelligent origin for life," *Diplomacy Post*, 3/22/14.

Earthworm or Draining Bathtub?

Psalm 22:6
"But I am a worm, and no man; a reproach of men, and despised of the people."

What creature lives deep underground, has a dark purple head and a blue-grey body, can grow up to nine feet in length and sounds like a bathtub draining of water? It's the Giant Gippsland earthworm of Australia, of course!

As these incredible creatures make their way through the soil, they shorten in length and then stretch their body to as long as fourteen feet. Despite their size, it is very difficult to study these giant earthworms because they never come to the surface unless flushed out by heavy rains. They prefer to spend their lives deep underground where they won't be cut to pieces by farmers tilling the ground.

"Burrows that are occupied by giant Gippsland earthworms have very wet walls," said Beverley D. Van Praagh, a biologist who has studied the creatures for over twenty years. "When the earthworms move quickly within their burrows, it makes a gurgling sound that is quite loud and can be heard above ground. The sound is a bit like water draining out of a bath and has been known to terrify the uninitiated."

Though most of us look at such creatures with disgust, nothing that God has made is without purpose. Earthworms perform a very valuable service in making the soil fertile. Life on Earth would not exist without the lowly earthworm! Eternal life, too, wouldn't exist were it not for Jesus, who is called a worm and was despised by the people He came to save.

Prayer: Heavenly Father, I thank You for Jesus, the lowly "worm" rejected by men but who is now exalted at Your right hand in heaven. Amen.

Ref: Matt Simon, "Absurd Creature of the Week: The 6-Foot Earthworm That Sounds Like a Draining Bathtub," *Wired* online, 3/28/14.

Dinosaurs Are Only Thousands of Years Old

Isaiah 43:20
"The beast of the field shall honour me, the dragons and the owls: because I give waters in the wilderness, and rivers in the desert, to give drink to my people, my chosen."

Previously, Creation Moments has reported on the surprising discovery of soft tissue, collagen, blood and even DNA in dinosaur bones. Such tissues shouldn't be able to survive for millions of years. Indeed, a team of researchers, using Carbon-14 dating on multiple samples of bone from eight dinosaurs, revealed that the bones were only 22,000 to 39,000 years old!

The ten-member research team presented their findings at the 2012 Western Pacific Geophysics Meeting in Singapore. After the applause died down and the conference came to a close, their report was quietly removed from the conference website by two chairmen who refused to accept the findings! Rather than challenge the data openly, they erased the report from public view without a word to the authors. Frankly, we're not surprised.

As John Michael Fischer of *NewGeology* rightly puts it, "Carbon-14 dates in the range of 22,000 to 39,000 years before [the] present, combined with the discovery of soft tissue in dinosaur bones, indicate that something is wrong with the conventional wisdom about dinosaurs."

Once again, we see how mistaken preconceptions can lead scientists astray. Since Carbon-14 dating is only used to date bones considered to be less than 50,000 years old, scientists never thought of using C-14 dating to determine the age of dinosaur bones. Or perhaps they were afraid of what they would find!

Prayer: Heavenly Father, I thank You for scientists who are bold enough to challenge conventional scientific wisdom. I pray that evolutionists keep running into stubborn facts that show them that Darwinism is wrong. Amen.

Ref: "Carbon-14-dated dinosaur bones are less than 40,000 years old," New Geology. Presentation video: https://www.youtube.com/watch?v=QbdH3l1UjPQ.

How Many Smells Can You Smell?

Psalm 45:8
"All thy garments smell of myrrh, and aloes, and cassia, out of the ivory palaces, whereby they have made thee glad."

Until recently it was commonly thought that humans can smell approximately 10,000 different odors. But that's no longer thought to be true. So how many different odors can your nose detect?

Scientists at Rockefeller University in New York set out to test the idea of 10,000 different odors. This number dates back to 1927, but it was never scientifically investigated. So they asked twenty-six people to identify a scent from three samples containing 128 molecules – two that were the same and one that was different. From this, they were able to tell how many different scents the average person would be able to distinguish if they were presented with all the possible mixtures that could be made from the 128 molecules.

So what was the official number? *One trillion different odors* – and even this, they say, is most likely an underestimate! Now, why would our Creator endow us with such an amazing sense of smell? Could He have been showing us how much He loves us? After all, the sense of smell enables us to appreciate the delightful aroma of flowers and food. In fact, we couldn't enjoy the food we eat if it weren't for our ability to smell.

Our sense of smell is one example of what is called God's common grace. He has given this gift to everyone, believers and unbelievers alike. But only believers can appreciate this as a gift from God and not an evolutionary accident of nature.

Prayer: Thank You, Lord, for being so generous in the gifts You have given us. Make me more appreciative of these gifts every time I sit down to a meal or smell a rose. Your blessings, while undeserved, are truly unlimited! Amen.

Ref: Helen Briggs, "Nose can detect one trillion odours," *BBC News Health*, 3/20/14.

Fishy Sign Language

Galatians 5:13
"For, brethren, ye have been called unto liberty; only use not liberty for an occasion to the flesh, but by love serve one another."

Humans aren't the only ones who use sign language, a recent study has found. Two types of fish are now known to use gestures to help them capture prey. Both the grouper and coral trout hunt cooperatively with other kinds of aquatic sea life. And they pull this off with sign language.

Though grouper are very fast in the open water, they are much too big to pursue prey that swim into small areas. But that's not a problem for the giant moray eel. Their slim body makes it easy to crawl into small holes. So when the grouper's prey flees to an area the grouper can't reach, it shows the eel where its lunch is hiding by rotating its body so that its head faces downward and then shakes its head back and forth in the direction of the out-of-reach meal.

Similarly, coral trout make gestures like the grouper's headstand signal to show their octopus hunting buddies the location of the prey. For the octopus, fitting into tight spaces is no problem.

We humans are often unable to accomplish a task by ourselves. If we are wise, we will ask others for help. And when others turn to us for help, we should do what we can to assist them. God didn't make us all alike. We need one another within the body of Christ. Use the gifts God has given you to be of service to others. And remain in fellowship with other believers so you can build up one another in Christ.

Prayer: Thank You, Father, for the gifts You have given me so that I may use them to serve others. And when I am in need of help, remove my pride so that I will allow others to minister to me. In Jesus' Name. Amen.

Ref: "Fish Use 'Sign Language' to Help Hunting Buddies," *Discovery News*, 4/9/13.

Fireflies Light Path to Brighter Bulbs

2 Corinthians 4:6
"For God, who commanded the light to shine out of darkness, hath shined in our hearts, to give the light of the knowledge of the glory of God in the face of Jesus Christ."

If you think that light-emitting diode bulbs are bright, wait until you see what scientists are developing – an LED that's much brighter than the bulbs we're already using. And we can thank one of God's most delightful creatures – the firefly – for lighting the path to these brighter bulbs!

An international team of researchers at the University of Namur in Belgium recently examined *Photuris* fireflies under a scanning electron microscope to see if they could identify the structures that produce the light. After identifying seven structures as candidates for increasing the brightness, computer simulations helped them determine it was the "misfit scales on the light-emitting organ's surface that boosted its brightness."

The multiple abrupt edges – much like the shingles on a roof – were seen to scatter light rays to increase the amount of light that escapes the firefly's glowing lantern. When a second team of scientists put this to the test by etching a similar shingle-like pattern on existing LED bulbs, they found it boosted the bulb's light output by 55 percent.

We don't know when these bulbs will be available to consumers. Perhaps they'll be on store shelves by the time you hear this. But we do know that these brighter LEDs didn't come about by accident. They were designed by scientists and engineers who knew where to look for answers. After all, engineers know a good design when they see one!

Prayer: Thank You, Lord, for the firefly and the many other marvelous creatures You made. I pray that my friends and family will see these designs and desire to learn more about their Designer. Amen.

Ref: M. Hoff, "Fireflies Inspire Brighter LEDs," *Discover*, 6/2/14.

Shooting at Evolution's Clay Pigeon

Isaiah 64:8
"But now, O LORD, thou art our father; we are the clay, and thou our potter; and we all are the work of thy hand."

Have you seen those animated films featuring characters made out of clay? Well, scientists at Cornell University have now published a study in which they claim that clay could have helped real life arise spontaneously from non-life millions of years ago. Scientists refer to this as *abiogenesis*. But Creation Moments says it's far more likely that an animated clay character could suddenly come to life.

The Cornell researchers claimed that clay was a key ingredient when life spontaneously emerged from non-life in Earth's early years. In other words, forget what Louis Pasteur accomplished in his experiments that proved life doesn't arise spontaneously from non-life. According to Cornell professor Dan Luo, "We propose that in early geological history, clay hydrogel provided a confinement function for biomolecules and biochemical reactions."

Not all scientists, however, agree with such speculations. For instance, Dr. Kevin Anderson, a research microbiologist with a Ph.D. in microbiology, said that despite the Cornell report, "there is not a shred of evidence that life can form spontaneously under any conditions."

Evolutionary biologists frequently attempt to break scientific laws – in this case, the law of biogenesis that states that life only comes from life. Let us remind evolutionists that biogenesis is not just a good idea, it's the law!

Prayer: Father, I pray that You will help people see the folly of believing in something as foolish as abiogenesis. I ask You to show them that life – and eternal life – come only from You. In Jesus' Name. Amen.

Ref: "Evolutionists Claim Clay Caused Life to Spontaneously Emerge From Non-Life," *Christian News*, 11/11/13.

Are Animals Just as Smart as Humans?

Genesis 1:27
"So God created man in his own image, in the image of God created he him; male and female created he them."

Over the years, Creation Moments has been introducing you to animals that are superior to humans when it comes to such things as eyesight, hearing and the ability to survive in conditions that are too hot or cold for us humans. But now, some scientists in Australia are saying that animals are *smarter* than us!

According to University of Adelaide experts in evolutionary biology, "Humans have been deceiving themselves for thousands of years that they're smarter than the rest of the animal kingdom, despite growing evidence to the contrary." The scientists have even written a book called *The Dynamic Human* that says that some animals may actually be brighter than we are.

According to Professor Maciej Henneberg, a professor of anthropological and comparative anatomy, "The animal world is much more complex than we give it credit for. Animals offer different kinds of intelligences which have been underrated due to humans' fixation on language and technology."

No, Professor Henneberg, creationists do not underrate the intelligence of animals. We have always known that God created each animal with gifts, abilities and the necessary intelligence to survive. What truly separates humans from the animals is not intelligence but the fact that God made us in His image. As it says in Genesis 2:7, "And the LORD God formed man of the dust of the ground, and breathed into his nostrils the breath of life; and man became a living soul."

Prayer: Thank You, Father, for setting us apart from every other living thing You created. I also thank You for giving us Your Son and Your Holy Spirit so that we can have eternal life. Amen.

Ref: D. Mosbergen, "Human Intelligence Isn't Superior To That Of Other Animals, Researchers Say," *The Huffington Post,* 12/12/13.

Black or White?

Zechariah 6:6
"The black horses which are therein go forth into the north country; and the white go forth after them; and the grisled go forth toward the south country."

Are zebras white with black stripes or are they black with white stripes? If you're thinking that zebras just have a coat of two colors and that they don't have a "background" color at all, think again. Embryologists tell us that the zebra's background color is black and that their stripes and bellies are white additions.

But scientists are now trying to find out why zebras have stripes at all. The best they can tell, the stripes may make zebras less appetizing to large biting flies that often carry fatal diseases. Using painted horses, Swedish scientists found that zebra stripes disrupt light patterns that tsetse flies and horseflies use to find food and water.

One of the authors of the study, evolutionary ecologist Susanne Åkesson, said the study is being used to show that the zebra's stripes are a result of evolution and natural selection.

As we've seen so many times before, the evolutionary ecologist will not see this as an example of an ingenious design by a Master Designer. Evolutionists are taught to believe in no design and no designer. Creationists, on the other hand, accept what common sense tells us. We can't help but see a most magnificent design in what God has created.

Since the Bible has never been proved wrong about anything, we can know for certain that zebras were painted, as it were, by an Artist who cares for all His creatures – especially us!

Prayer: Thank You, Jesus, for creating the zebra, a creature that reminds me of the Bible verse: "And with his stripes we are healed." Just as You provided for the zebra in this life, so have You provided for me in the life to come. Amen.

Ref: J.J. Lee, "Mystery of Zebra's Stripes Finally Solved?" *Science*, 2/9/12.

See Dick and Jane Evolve

Proverbs 22:6
"Train up a child in the way he should go: and when he is old, he will not depart from it."

In his book, *The Blind Watchmaker,* evolutionary evangelist Richard Dawkins famously declared: "Biology is the study of complicated things that give the appearance of having been designed for a purpose." Yes, seeing design in nature is so natural, so common-sensical that even young children innately know that nature is filled with purposeful design.

Of course, evolutionists can't sit back and let common sense persuade children to accept biblical creation. So Boston University psychologist Deborah Kelemen and her team set out to prove it's possible with Darwinian storytelling to make children believe in natural selection rather than a Creator God. After giving children (ages 5-8) picture books that illustrated an example of natural selection, many of the children agreed with evolution.

Even *The Wall Street Journal* was pleased: "These results do suggest that simple story books like these could be powerful intellectual tools. The secret may be to reach children with the right theory before the wrong one is too firmly in place." The study, they said, "also suggests that we should teach children the theory of natural selection while they are still in kindergarten instead of waiting, as we do now, until they are teenagers."

The success of this study leads us to conclude that books like these will be used against you and your children in the future. That's why you need to take a very active role in your children's education. Review their textbooks. Ask questions. Attend school board meetings. And if things get too bad, consider alternatives to public education.

Prayer: Thank You, Father, for my children. I pray that You will protect them from those who desire to cause them harm. In Jesus' Name. Amen.

Ref: "Young Children Can Be Taught Basic Natural Selection Using a Picture-Storybook Intervention," *Psychological Science,* April 2014 25: 893-902, first published on 2/6/14. "See Jane Evolve: Picture Books Explain Darwin," *The Wall Street Journal,* 4/18/14.

Scientists Build the Universe

Genesis 1:1
"In the beginning God created the heaven and the earth."

After three months of number crunching, using 8,000 computers running in parallel, scientists say they have been able to build a model of the stars and galaxies of our "evolving universe" in a computer simulation.

The simulated universe is called Illustris. It not only shows what the universe looks like now, it shows what it looked like in the past. Discovery.com notes that the simulation "only uses equations from theories constructed from decades (even centuries) of astronomical observations and allowed to evolve with time. The result," they said, "is nothing short of breathtaking and it can be hard to distinguish the model from real observations."

Even so, the scientists behind Illustris admit there are "anomalies in the simulation that don't match our observations." This should come as no surprise to creationists. After all, how could a computer simulation based on "equations from theories" match what we find in the real universe? How sad that the scientists claim the false premise of evolution for their work and spend so much time trying to show how the universe began without God. God's fingerprints can be clearly seen all over the creation – especially in the stars, nebulae and galaxies He created on the fourth day of Creation Week.

If scientists spent some of their time searching the Scriptures rather than the starry skies, they might develop an even better simulation. In fact, they might even discover the One who created the universe and died so that they might have everlasting life.

Prayer: Father, I pray for the work of creationists involved in the sciences – that they would make significant scientific contributions that cannot be ignored by scoffers and Bible skeptics. In Jesus' Name. Amen.

Ref: Ian O'Neill, "Mind-Blowing Computer Simulation Recreates Our Universe," *Discovery News*, 5/7/14.

Bad News, Worse News, Good News

Deuteronomy 6:6-7a
"And these words, which I command thee this day, shall be in thine heart: And thou shalt teach them diligently unto thy children..."

According to the latest Gallup poll, four in ten Americans believe God created the Earth and anatomically modern humans less than 10,000 years ago. When you consider the monopoly on government and education held by the evolution lobby, that's not really so bad. Gallup also finds that half of Americans now believe humans evolved over millions of years, though most of these people say that God guided the process.

"Religious, less educated, and older respondents were likelier to espouse a young earth creationist view – that life was created some 6,000 to 10,000 years ago," according to the Gallup Values and Beliefs Survey. Though the percentage of people who believe in creationism has changed little over the decades, the bad news is that *the percentage of people who believe humans evolved without God has more than doubled!*

And college only makes things worse. While 57 percent of students who received just a high-school education are creationists, only 25 percent of those with a college degree believe in creation. Evolution's stranglehold on the minds of young people should be a grave concern for all Christians.

But there's good news to report. To the credit of some churches, more than two-thirds of those who attend weekly religious services believe in a young Earth, compared with just 23 percent of those who never go to church. We would venture to say that this is due in part to churches that give their youth a solid foundation in the Bible and the truth of biblical creation. Is your church one of them?

Prayer: Father, I pray that You will help parents and churches prepare their children to stand strong against the false teaching they will receive in school. Amen.

Ref: Tia Ghose, "4 in 10 Americans Believe God Created Earth 10,000 Years Ago," *LiveScience*, 6/6/14.

Where Did All the Water Go?

Genesis 8:3
"And the waters returned from off the earth continually: and after the end of the hundred and fifty days the waters were abated."

Evolutionists and atheists like to think they've got Bible-believing creationists stumped when they ask, "If a worldwide flood really happened, where did all the water go? If Noah's flood covered the mountaintops, what happened to all that water?"

Creationists have often responded by saying that the water ended up in the deep ocean trenches like the Mariana trench in the western Pacific. The trench is so deep, the world's tallest mountains could be submerged without breaking the ocean's surface!

But another answer was suggested by the recent discovery of a vast reservoir of water that's three times larger than all of the world's oceans combined! As reported in *NewScientist*, "The water is hidden inside a blue rock called ringwoodite that lies 700 kilometres – that's 434 miles – underground in the mantle, the layer of hot rock between Earth's surface and its core."

According to researcher Steven Jacobsen of Northwestern University, "It's good evidence the Earth's water came from within." This statement by a secular scientist agrees with the Bible when it says that in addition to the rain, the "fountains of the great deep" were broken up.

This could also help explain where all the water went after the flood subsided. It suggests that the waters of the flood ended up in the sponge-like ringwoodite that's buried beneath the entire United States and possibly the whole world. Christians never need to fear the charges brought against the Bible by atheists and scoffers.

Prayer: Father, thank You for the many scientific discoveries that confirm the historicity of the Bible. In Jesus' Name. Amen.

Ref: A. Coghlan, "Massive 'ocean' discovered towards Earth's core," *NewScientist*, 6/12/14. *Science* 13 June 2014: Vol. 344 no. 6189 pp. 1265-1268 DOI: 10.1126/science.1253358.

Universe Not Expanding After All?

Isaiah 40:22
"It is he that sitteth upon the circle of the earth, and the inhabitants thereof are as grasshoppers; that stretcheth out the heavens as a curtain, and spreadeth them out as a tent to dwell in:"

If we can learn anything from science, it is this – the scientific facts of today won't necessarily be the facts of tomorrow. The simple truth is that a scientific fact that is no longer a fact was never a fact to begin with!

Take, for example, the so-called scientific fact of the expanding universe. We've all been told that this is a scientific fact, a relic of the Big Bang that explains the red shift of stars as they speed away from us. But now, a team of astrophysicists from Lawrenceville Plasma Physics is calling that well-known fact into question.

According to Sci-News.com, the results of their tests "are consistent with what would be expected from ordinary geometry if the Universe was not expanding, and are in contradiction with the drastic dimming of surface brightness predicted by the expanding universe hypothesis." In other words, this "new evidence, based on detailed measurements of the size and brightness of hundreds of galaxies, indicates that the universe is not expanding after all."

We'll let scientists battle it out over whether the universe is expanding. Their views are bound to change anyway. But if the universe is expanding, creationists can point to the fact that God *stretched out* the heavens as a curtain during Creation Week.

Prayer: Father, thank You for scientists who increase our understanding of the universe You created. But never let me forget that man's knowledge falls far short of the knowledge we find in the pages of Scripture. Amen.

Ref: Eric J. Lerner et al. UV surface brightness of galaxies from the local Universe to $z \sim 5$. Int. J. Mod. Phys. D, published online May 02, 2014; doi: 10.1142/S0218271814500588. "Universe Is Not Expanding After All," Sci-News.com. 5/23/14.

Dung Beetle Navigation

Job 37:14
"Hearken unto this, O Job: stand still, and consider the wondrous works of God."

Humans, birds and seals are all known to navigate by the stars. But now, the lowly dung beetle could be the first example of an insect being a stargazer. These lowly insects appear to use the stars of the Milky Way to know where they are going!

Dung beetles, you see, like to move in straight lines. When they come across a pile of droppings, they make a small ball of dung and start pushing it away to a safe distance where they can dine alone, far away from their hungry neighbors. If they didn't know how to push it in a straight line, they would run the risk of circling back to the original pile of droppings and having to fight off other dung beetles.

Dr. Marie Dacke of Lund University in Sweden had previously shown that dung beetles were able to travel in a straight line by taking cues from the Sun, the Moon, and even the pattern of polarized light formed around these light sources. But it was their capacity to maintain course even on clear, moonless nights that led to the new discovery.

Evolution, of course, is powerless to explain how the dung beetle can navigate by the stars – something that most people can't do without the help of technology. God has given them a set of instructions – or instincts – that enable them to survive. If God cares so much for the lowly dung beetle, just think about how much more He cares for you!

Prayer: Oh Lord, when I consider the unique abilities You have given to even the lowliest of creatures, I am in awe of Your unbounded love. Thank You for the gifts You have given me, especially the gift of eternal life. Amen.

Ref: Jonathan Amos, "Dung beetles guided by Milky Way," *BBC News,* 1/24/13.

A Fly with Ants on Its Wings

Romans 1:20
"For the invisible things of him from the creation of the world are clearly seen, being understood by the things that are made, even his eternal power and Godhead; so that they are without excuse."

When most people look at a photo of the *G tridens* fruit fly, they think it must have been created by a talented Photoshop artist. This fly – with images of ant-like insects on its wings – is no accident of nature. The images were put there by a Master Designer.

Dr. Brigitte Howarth, the fly specialist at Zayed University in the UAE who first discovered *G tridens*, said that the image on the wings is absolutely perfect, adding that a closer examination of the transparent wings reveals a piece of "evolutionary art." Each wing carries a precisely detailed image of an ant-like insect, complete with six legs, two antennae, a head, thorax and tapered abdomen. When threatened, the fly flashes its wings to give the appearance of ants walking back and forth. This confuses the predator, allowing the fly to escape.

As expected, the story and pictures of *G tridens* were soon flying all over the Internet. Evolutionists rejoiced, claiming that this fruit fly was a beautiful example of natural selection. But we have to point out that there is no evidence to support that claim. As usual, evolutionists simply assume that evolution is true. What they fail to do is show how those images of ants actually got on the fly's wings.

Creationists don't need to provide a step-by-step explanation of how these extraordinary wings developed. The fact is, the design is in their genes. God designed them that way!

Prayer: Heavenly Father, when I look at the things You created, I see what unbelievers are unable to see. Open their eyes so they, too, might see that You are the Designer who alone has the power to save them from their sins. Amen.

Ref: Anna Zacharias, "Fruit fly with the wings of beauty," *TheNational*, 6/28/13.

Maxwell's Prayer

Psalm 90:12
"So teach us to number our days, that we may apply our hearts unto wisdom."

The great scientist James Clerk Maxwell, a contemporary of Charles Darwin, was certainly no friend of Darwin's theory of evolution. Even so, the staunchest evolutionists today would tell you that Maxwell was a scientist of gigantic proportions, ranking right up there with celebrated scientists like Sir Isaac Newton.

In a previously aired Creation Moments broadcast, we told you that Maxwell believed that Jesus Christ is the Savior who came to deliver humanity from the results of sin. We also mentioned that a writing of his, found after his death, stated that the motivation for his work was that God had created all things just as Genesis says. And since God created humans in His image, scientific study is a fit activity for one's lifework.

But Maxwell's heart for God is expressed most clearly in a prayer he wrote that was found in his notes after he died in 1879 at the young age of 48. The rest of today's broadcast is in Maxwell's own words:

"Almighty God, Who hast created man in Thine own image, and made him a living soul that he might seek after Thee, and have dominion over Thy creatures, teach us to study the works of Thy hands, that we may subdue the earth to our use, and strengthen the reason for Thy service; so to receive Thy blessed Word, that we may believe on Him Whom Thou hast sent, to give us the knowledge of salvation and the remission of our sins. All of which we ask in the name of the same Jesus Christ, our Lord."

Prayer: Heavenly Father, I pray that You would give more scientists the faith that guided men like James Clerk Maxwell. Help them to see science as studying the works of Your hands. In Jesus' Name. Amen.

Ref: http://www.doesgodexist.org/JulAug13/Maxwell--Scientists.html.

Vanilla Stumps Evolutionists

Genesis 1:11
"And God said, Let the earth bring forth grass, the herb yielding seed, and the fruit tree yielding fruit after his kind, whose seed is in itself, upon the earth: and it was so."

The next time you enjoy vanilla ice cream, keep in mind that vanilla wouldn't even exist if evolution were true. Evolutionists don't have a blooming idea about the origin of a special symbiotic relationship between a particular flower and a very special bee.

You see, vanilla comes from the *vanilla planifolia* plant which develops into the Mexican vanilla orchid. Unlike most orchids, this one blooms only one morning each year. The orchid also has a hood-like membrane that covers the part that produces pollen. These two facts make pollination nearly impossible.

But the God who created this plant also created the Mexican Melipona Bee – the only insect that knows how to pollinate the orchid. After landing on the flower, the bee lifts up the hood, collects the pollen and then flies off to another flower. Once pollinated, the orchid produces a vanilla bean. If not pollinated within eight to twelve hours, the flower wilts and drops from the mother vine.

Without the Mexican Melipona Bee, you wouldn't be able to enjoy any of the delicious foods made with vanilla extract. So we would ask evolutionists – which came first: the orchid or the bee? And how did the bee learn to pollinate the vanilla orchid?

At best, evolutionists can only offer guesses. But creationists know that the bee and the flower are a match made in heaven. They enjoy a symbiotic relationship so that each can survive ... and so we can enjoy the tasty fruit of their labor!

Prayer: Father, thank You for filling our planet with so many foods that not only nourish us but bring us pleasure. In Jesus' Name. Amen.

Ref: Brett Petrillo, "Vanilla Ice Cream Defies Evolution," BP's Fuel for Thought, 1/28/14.
http://bioweb.uwlax.edu/bio203/s2009/ruud_kirs/Life%20History%20-%20Reproduction.htm.

Feathered Dinosaurs

Isaiah 31:5
"As birds flying, so will the LORD of hosts defend Jerusalem; defending also he will deliver it; and passing over he will preserve it."

A new species of dinosaur has been found. Scientists are calling *Changyuraptor yangi* the biggest feathered dinosaur ever discovered. Here at Creation Moments, however, we're calling it a feathered bird. Dinosaur-to-bird evolution is nothing more than a flight of fancy.

Naturally, USA Today, BBC News any many other news outlets are calling it a four-winged dinosaur. *Changyuraptor yangi*, they say, is a new species of dinosaur that offers "clues to the origin of flight – and the transition from feathered dinosaurs to birds." Research scientists also say that "the new fossil possesses the longest known feathers for any non-avian dinosaur."

Non-avian? No, we're still calling them birds, and here's why. For one thing, dinosaurs and other reptiles have scales, which are folds in their skin. Birds, on the other hand, have feathers which grow out of follicles. Scales and feathers are completely different. No known fossils, in fact, provide evidence of a transition from scales to feathers.

Furthermore, for a dinosaur to evolve into a bird, it would need to develop hollow bones, it would have to gain powerful flight muscles and develop a new heart with four chambers rather than three.

So, too, do we need a new heart to understand the things of God. We don't evolve our old heart into a new one. Our new heart comes to us only through the work of the Holy Spirit.

Prayer: Thank You, Father, for sending the Holy Spirit so we can become new creatures that have a close and enduring relationship with You. Amen.

Ref: "Scientists discover largest four-winged dinosaur to date," *USA Today*, 7/15/14. "Four-winged dinosaur is 'biggest ever'," *BBC News Science & Environment*, 7/16/14.

The World's Ugliest Animal?

Romans 5:8
"But God commendeth his love toward us, in that, while we were yet sinners, Christ died for us."

When this fish is taken out of the water, its face almost looks like a very sad person. In 2013 it was voted the "World's Ugliest Animal." What is this creature that was adopted as the mascot of the Ugly Animal Preservation Society? It's the appropriately named blobfish.

Photos of the ugly blobfish have been making the rounds on the Internet. And yet, even the blobfish has beauty when you consider that it was designed to function perfectly in its environment.

Blobfish inhabit the deep waters off the coasts of Australia, Tasmania and New Zealand. If you wanted to see one in its natural habitat, you would have to dive to a depth of between 2,000 and 3,900 feet where the pressure is several dozen times higher than at sea level. The pressure is no problem for the blobfish, though. Its jelly-like body is slightly less dense than the cold saltwater, allowing it to float effortlessly just above the sea floor.

Blobfish don't have much muscle for swimming, but they don't need it. They simply swallow edible matter that floats into their mouth. Sadly, they are an endangered species because of fishing trawlers dragging their nets on the seafloor.

Yes, the blobfish is far from attractive. But it serves as a reminder that sinful human beings are unattractive in the sight of God. And yet, while we were still sinners, God sent His Son to die for us and to exchange His righteousness for our sins, making us acceptable in God's sight. What a Savior!

Prayer: Father, thank You for sending Your Holy Spirit to reveal the ugliness brought about by sin. Thank You also for sending Your Son to remove my sin and make me righteous in Your sight. Amen.

Ref: "So you think you've had a bad day? Spare a thought for the world's most miserable-looking fish, which is now in danger of being wiped out," *Daily Mail*, 1/26/10.

The Little Dipper

Psalm 86:9-10
"All nations whom thou hast made shall come and worship before thee, O Lord; and shall glorify thy name. For thou art great, and doest wondrous things: thou art God alone."

Also known as the water ouzel, the dipper is a small bird so named because of its characteristic dipping or bobbing motion when perched beside the water of fast-flowing rivers.

As author Douglas Sharp notes in *Revolution Against Evolution,* the dipper "not only flies in the air, but swims underwater with his wings. He also strolls on the bottom of the stream, overturning rocks with his beak and toes to feed on various water creatures. Air sacs provide buoyancy, enabling him to rise to the surface. He 'blows his tanks' to submerge. Since he does not have webbed feet, he uses his wings as underwater oars."

The author then asks evolutionists: "How many eons of diving school did this bird endure before he mastered the delicate balance of the air and water environments? These unique air sacs will either work, or they won't. These functions would have to be perfected before our skinny-dipping friend would ever discover the juicy morsels on the bottom of the stream."

Wikipedia, which is certainly no friend to biblical creation, mentions two more design features of this amazing little bird. The dipper's eyes have well-developed focus muscles that can change the curvature of the lens to enhance underwater vision. Dippers also have nasal flaps to prevent water from entering their nostrils. What did they do before evolution produced those flaps? Choke to death?

Obviously, evolutionists cannot account for this little dipper's many design features, but creationists can!

Prayer: Heavenly Father, thank You for the diversity I find in Your creation. Each new plant or animal I learn about fills me with awe of Your infinite creativity! Amen.

Ref: http://www.rae.org/pdf/rae.pdf. *Revolution Against Evolution,* 2013, Douglas Sharp, p. 49. Third edition. Decapolis Books. http://en.wikipedia.org/wiki/Water_ouzel.

Planets Vanish!

Proverbs 14:25
"A true witness delivereth souls: but a deceitful witness speaketh lies."

In light of their expectation to find life on other planets, the news story I'm going to tell you about today must have been sheer torture for *ScienceNews* magazine to report.

Here's what they wrote: "Two planets considered among the most promising for life outside the solar system don't exist.... The signals embedded in starlight that were attributed to the planets may instead have been caused by the changing magnetic activity of their star, Gliese 581."

Scientists say that the error reinforces the need for "meticulous analyses to separate planets' signals from those generated by spots and flares on stars." Stéphane Udry, an astronomer at the University of Geneva, agrees: "This is a big warning concerning the interpretation of [small] signals as being planets."

In light of these "vanishing" planets, one has to wonder how many other planets are, in fact, non-existent. Could many or most of the millions of planets that are now thought to be hospitable to life be nothing but spots and flares on stars?

Scientists pride themselves on correcting their errors, and this is a good example of that. But when we think of all of the so-called evidences for evolution still in school textbooks, it is abundantly clear that scientists still have much to learn about the errors they stubbornly refuse to correct. Creationists can rejoice that, unlike science textbooks, there is one book that never needs to be corrected or updated – the Bible!

Prayer: Heavenly Father, I pray for scientists who don't know You, the Source of all truth. Many of them are filled with pride that prevents them from gaining eternal life. Touch them with Your life-giving Spirit, I pray. Amen.

Ref: "Exoplanets once trumpeted as life-friendly may not exist," *ScienceNews*, 7/3/14. Magazine issue: August 9, 2014.

World's Best Smelling Animal

Psalm 45:8
"All thy garments smell of myrrh, and aloes, and cassia, out of the ivory palaces, whereby they have made thee glad."

What animal is the world's best smeller? Scientists now think it's the African bush elephant. As we've mentioned previously, elephants possess a versatile trunk that can toss logs, grasp food, spray water and even pick up branches to use as tools. But new research reported in *Genome Research* and *ScienceNews* has found that these magnificent creatures carry about 2,000 genes for olfactory receptors. That's about five times more olfactory genes than we have. It's also a lot more than renowned sniffers like rats, which have only 1,200 olfactory receptor genes. Dogs have even fewer – about 800.

Just how well can an African bush elephant smell? They are actually able to distinguish between the Maasai, an ethnic Kenyan group that hunts elephants, from the Kamba, who are farmers. Smelling the difference between these two groups causes the elephant to respond accordingly.

As can be expected, the research team attributed the elephant's superior sense of smell to evolution. They said that the original smell-sensing gene must have duplicated in elephants as mammal species diverged from one another. You see, evolution is their assumption; it is not proven from testing.

We see no evidence whatsoever that the elephant's sense of smell has anything to do with evolution. A far more reasonable explanation is that they were given this gift by their Creator who knew exactly what they would need to survive. Yes, that same Creator who has given us what we need to survive the punishment that we deserve when Jesus was crucified!

Prayer: Lord, I am filled with awe when I think about the creatures You have made. From the mighty elephant to the lowly worm, all were made by You! Amen.

Ref: Y. Niimura, A. Matsui, and K. Touhara, "Extreme expansion of the olfactory receptor gene repertoire in African elephants and evolutionary dynamics of orthologous gene groups in 13 placental mammals," *Genome Research*. Published online: 7/22/14. "Elephant's big nose wins most sensitive sniffer," *ScienceNews*, 7/22/14.

The Purple Thief

Psalm 111:4
"He hath made his wonderful works to be remembered: the LORD is gracious and full of compassion."

Birds and insects that take nectar from a flower without picking up any pollen are known as nectar robbers. Now, you'd probably think that nectar robbers would be harmful to plants and trees, but the desert teak tree couldn't survive without a nectar robber – the purple sunbird.

In order to reproduce, this tree needs birds to pollinate its flowers. But since a tree cannot reproduce with its own pollen, it needs birds to fly from flower to flower and from one tree to another. Anything that encourages the pollinating birds to fly farther away helps out the teak trees.

That's where the purple thief comes in. Researchers at the University of Delhi discovered that the purple sunbird visits the flowers one hour before the pollinating birds arrive. The sunbird has a long, sharp beak that pierces the base of the flower to feed, so it doesn't pick up any pollen. It does, however, empty the flower of about 60 percent of its nectar, leaving relatively little for the pollinators. This means that the pollinators will have to travel to more flowers and trees to get enough food, spreading pollen wherever they stop for a meal.

The researchers noted that "the robber plays a constructive and crucial role in the reproductive performance of [a] threatened tree species." How right they are. And this unusual but crucial dining arrangement shows once again what an ingenious God we serve!

Prayer: Lord, only You could come up with such an ingenious way to help the desert teak tree to reproduce! Surely such an arrangement could not have come about through blind chance! Amen.

Ref: "These trees don't mind getting robbed," *ScienceNews*, 7/25/14.

Scientist Fired for Dinosaur Discovery

1 John 4:6
"We are of God: he that knoweth God heareth us; he that is not of God heareth not us. Hereby know we the spirit of truth, and the spirit of error."

If there is an unwritten law in the field of science, it is this: Thou shalt not discover anything that even suggests that evolution could be wrong. And a second unwritten law is like unto it: If you *do* discover anything that undermines evolution, keep it to yourself.

A scientist was fired from his job at California State University, Northridge, after discovering fossil evidence that supports a young Earth and then publishing his findings. While at a dig at Hell Creek formation in Montana, scientist Mark Armitage came upon the largest triceratops horn ever unearthed at the site. When he examined the horn under a high-powered microscope, he was shocked to see soft tissue. This discovery stunned other scientists because it indicated that dinosaurs roamed the Earth only thousands of years ago rather than 60 million years ago.

In his wrongful termination and religious discrimination lawsuit, court documents revealed that a university official challenged his motives by shouting at him, "We are not going to tolerate your religion in this department!" Armitage joins a growing number of scientists who have lost their jobs for challenging the sacred cow of evolution. Many institutions of higher learning are no longer interested in pursuing the evidence wherever it may lead. They only care about evidence that leads them straight to their foregone conclusion that evolution is a fact.

Creationists need not fear any new scientific discovery because nothing can successfully contradict the Word of God!

Prayer: Father, I pray that You would give Christians who are involved in the sciences courage and wisdom in a field that is increasingly hostile to their beliefs. Amen.

Ref: http://www.pacificjustice.org/press-releases/university-silences-scientist-after-dinosaur-discovery. "University Silences Scientist After Dinosaur Discovery," 7/23/14. Pacific Justice Institute is a non-profit legal organization dedicated to defending religious, parental, and other constitutional rights.

Evolution Produces Crybabies

Ruth 4:16
"And Naomi took the child, and laid it in her bosom, and became nurse unto it."

If you're a parent, you undoubtedly remember being awakened by your little angel crying in the middle of the night. Sometimes it's dad who gets out of bed to see what's wrong. But more often, it's mom who gets up to breastfeed her baby. But now a Harvard scientist tells us that the baby who demands a nighttime meal has a more sinister reason. The child is trying to prevent siblings from being born.

Does that make sense to you? It does to evolutionary biologist David Haig. Writing in *Evolution, Medicine and Public Health,* he suggested that if its parents had another baby, this would mean having to share mom and dad. So babies are "programmed" to do everything they can to keep this from happening. In our past, Haig proposed, babies who cried to be nursed at night had a survival edge.

Oh really? If the baby is trying to prevent mom and dad from having more babies, doesn't this go against the parents' "evolutionary" drive to bear the greatest number of children to benefit the species?

Evolutionists who dream up such theories are rewarded with appearances on national newscasts. They also arouse the interest of other scientists and see their papers published in respected scientific journals. But most of all, they are praised for suggesting once again that everything in our world rests upon the foundation of evolution.

Thankfully, Bible-believing scientists have a far different foundation – the Lord Jesus Christ!

Prayer: Heavenly Father, is it too difficult for scientists to understand that babies cry at night because they are hungry or want to be held? As Your child, I am grateful that You are there to comfort me at any time of day or night. Amen.

Ref: L. Sanders, "Babies cry at night to prevent siblings, scientist suggests," *ScienceNews,* 4/22/14. "Troubled sleep, Night waking, breastfeeding and parent–offspring conflict," D. Haig, *Evolution, Medicine and Public Health*, Volume 2014, Issue 1, pp. 32-39.

Humans with Tails?

1 Peter 3:15
"But sanctify the Lord God in your hearts: and be ready always to give an answer to every man that asketh you a reason of the hope that is in you with meekness and fear:"

Creation Moments has aired a number of programs about theistic evolution. We have also posted a number of articles at our website about this false and dangerous belief. We interact with theistic evolutionists almost every day and have found them to be almost identical to atheists in their exceedingly low view of the Bible.

Not long ago, a theistic evolutionist revealed once again how far they will go to defend evolution. During a debate with the Discovery Institute's Dr. Stephen Meyer, theistic evolutionist Dr. Karl Giberson showed a photo of a human infant with a monkey-like tail. He did this to drive home his point that humans share a common descent from a tailed ancestor.

The only thing Dr. Giberson failed to mention was that he had found the photo at the website Cracked.com, an online humor website inspired by *Mad* magazine. When the source of the photo was revealed several days later, Dr. Giberson apologized, promising to never use the photo again. But like many apologies from evolutionists, this one fell a little flat. He protested: "No point was implied by the image beyond what is well established by other legitimate images. At no point did I imply that the image I showed was evidence."

Christians should never even think of being deceptive when sharing the truth of biblical creation. Yes, we may make a mistake. But when we do, a sincere apology might make all the difference in an unbeliever's life.

Prayer: Heavenly Father, You desire that we be people of integrity and honesty. Lord, use any means to remove my pride so I will admit my mistakes and correct them. Amen.

Ref: "Karl Giberson Apologizes for Photoshopped Image of Tailed Baby," *Evolution News and Views*, 7/17/14.

The "Invisible" Mouse

Isaiah 1:18
"Come now, and let us reason together, saith the LORD: though your sins be as scarlet, they shall be as white as snow; though they be red like crimson, they shall be as wool."

It looks like H.G. Wells' *The Invisible Man* has just come closer to becoming a reality. A group of Caltech researchers recently announced they have succeeded in making a mouse transparent. The researchers said that their new scientific technique – called Perfusion-assisted Agent Release in Situ, or PARS for short – will help scientists study organs and tissues in the lab, and could even help diagnose illnesses in humans.

So how did they pull this off? They did it by pumping a detergent through the dead mouse's tissues to remove the fat cells. Sure enough, this made the mouse transparent. However, it also caused the mouse to quickly decompose. To solve this problem, the researchers replaced the fatty lipids with a transparent gel to maintain the mouse's body structure. By pumping this cocktail of detergent and gel through the mouse's circulatory system, they made the mouse's body transparent and ready for research.

According to *The Washington Post,* labs have already begun using the lipid-clearing technique on tissue from human biopsies. Using the technique for finding cancerous cells, said one researcher, is a no-brainer.

Truly, we live in an age when it seems as if science can solve any problem. But there's one big problem that science will never be able to solve – the problem of sin and death. Only Jesus can take a person's sin-stained heart and turn it white as snow. Only Jesus can make our sins vanish from God's sight!

Prayer: Heavenly Father, I am thankful to live at a time when medical advances are improving the quality of human life. I thank You most of all for eternal life through Jesus Christ, my Lord! Amen.

Ref: R. Feltman, "Why a see-through mouse is a big deal for scientists," 7/31/14. "Single-Cell Phenotyping within Transparent Intact Tissue through Whole-Body Clearing," *Cell*, 5/6/14.

Tisha B'Av

Matthew 24:1-2
"And Jesus went out, and departed from the temple: and his disciples came to him for to shew him the buildings of the temple. And Jesus said unto them, See ye not all these things? verily I say unto you, There shall not be left here one stone upon another, that shall not be thrown down."

One of the most solemn days of the year for the Jewish people is not a biblical holy day. No, this national day of mourning – known as Tisha B'Av – is commemorated because of all the catastrophes that happened to Israel throughout history on the ninth day of the month of Av.

According to rabbinic tradition, five disastrous events took place on the ninth of Av. We have time to mention just three. It was on the ninth day of Av in the year 587 BC that King Solomon's Temple was destroyed by Nebuchadnezzar and the Babylonians after a two-year siege. The people of Judah were then sent into Babylonian exile.

Nearly seven hundred years later – again on the ninth of Av – the Second Temple was destroyed by the Romans in the year 70 AD. This date also marks the beginning of the Jewish exile from the land of their fathers, an exile that continued for almost nineteen hundred years.

The third event we will mention today happened in the year 132 AD. It was on the ninth day of Av that the Romans crushed the Bar Kokhba revolt, killing more than 100,000 Jews and ending the Jewish hope of overthrowing Roman rule. Since many Jews believed Bar Kokhba to be the long-awaited Messiah, this was a tragedy of epic proportions.

Even more tragic is that so many Jews back then and today do not realize that the Messiah has already come ... and is coming again! Tell every Jewish person you know that their Messiah is Jesus!

Prayer: Father, I pray for the Jewish people. Never let me forget that Your Son came to us through the line of Judah to be the sacrifice for our sins. Amen.

Ref: Wikipedia entry on "Tisha B'av."

"That's Evolution for You!"

Luke 12:6-7
"Are not five sparrows sold for two farthings, and not one of them is forgotten before God? But even the very hairs of your head are all numbered. Fear not therefore: ye are of more value than many sparrows."

"That's evolution for you!" This is what Dr. Rick Fairhurst, malaria researcher at the National Institute of Allergy and Infectious Diseases, said when talking about malaria parasites that are becoming resistant to drugs that had once been effective at fighting the deadly disease.

The World Health Organization estimates there were 207 million cases of malaria worldwide in 2012, killing 627,000 people – most of them children under age five. *ScienceNews* laments that there is no approved vaccine against the protozoan malaria parasite, which is spread by mosquitoes and causes fever, chills, convulsions and more severe symptoms.

The whole point of the article is that "the history of malaria treatment is peppered with drugs to which the parasite has become resistant." But as Creation Moments has pointed out many times in the past, drug-resistant bacteria didn't get that way through evolution. Drug-resistant bacteria represent a *loss* of information. It would be like saying that you evolved resistance to tennis elbow because you lost both of your arms in a car accident.

To evolutionists, though, everything is an example of evolution. This is nothing but shoddy thinking resulting from what they already believe to be true. It is the same kind of bad thinking that led to the banning of DDT, which – when sprayed in small quantities in homes – is a harmless pesticide that kills parasite-carrying mosquitoes. DDT could have prevented the deaths of millions. It is no less tragic when scientists say "That's evolution for you!" when it has absolutely *nothing* to do with evolution!

Prayer: Father, I pray that world leaders will realize that human life is of infinitely more worth than imagined threats to our environment. Amen.

Ref: "Resistance to key malaria drug spreads," *ScienceNews* 7/31/14.

Digging Up the Dirt on Fraudulent Fossils

1 Samuel 12:24
"Only fear the LORD, and serve him in truth with all your heart: for consider how great things he hath done for you."

How far will scientists and museums go to get people to believe in the theory of evolution? As scientist Carl Werner reveals in his book and on his webpage *Evolution: The Grand Experiment,* they will readily fabricate fake fossils and pass them off as real.

Dr. Werner said, "Whales with four legs, walking on land, are currently considered one of the best fossil proofs of evolution." But after interviewing the two scientists who reconstructed the fossils of the three famous walking whales – Rodhocetus, Pakicetus and Ambulocetus – he concluded that the scientists had created false models of these skeletons and skulls and passed them off as real to the biggest and most respected museums in the world.

In a Generations Radio interview, he told listeners, "If you go to the largest museums in the world right now, there's fabricated fossil after fabricated fossil, where they've attached whale body parts to walking animals and said, 'Oh, look, we have walking whales.' These are on display currently, and I have interviewed the scientists who put the whale body parts onto four-legged animals, and they admitted to it."

Dr. Werner predicts that someday the fossil fakes will be quietly removed from museums. But for now, these fakes will deceive those who go to museums and websites where they will learn about "whales with legs." This is yet another reason why those of us with the truth must not remain silent but must share the truth with all who will listen.

Prayer: Heavenly Father, telling evolutionists about the fraudulent fossils in museums can be important, but in the end I know that the only thing that can change their heart is the Gospel. Amen.

Ref: http://thegrandexperiment.com/whale-evolution.html. Press release dated 4/7/14.
http://www.sermonaudio.com/sermoninfo.asp?SID=513141026269. "Why Evolutionists Fake Evidence - Walking Whales - Another Fraud," Generations Radio, 5/13/14.

Penguins with Sunglasses

Hebrews 1:3
"Who being the brightness of his glory, and the express image of his person, and upholding all things by the word of his power, when he had by himself purged our sins, sat down on the right hand of the Majesty on high;"

Have you ever gone outside on a bright, sunny day and been almost blinded by the light? Then imagine what it must be like for penguins to go about their lives with the intense glare of polar sunlight reflecting off a snowy or watery landscape.

Penguins, of course, can't slip on a pair of sunglasses, but they don't need to. These marvelous birds have an external orange eye fluid that filters out blue and ultraviolet wavelengths from the solar spectrum. This gives them clear vision while protecting their eyes from harm. As you might have guessed, eagles, falcons, hawks and other birds of prey also have this fluid.

Inspired by these birds, scientists have developed an orange-colored dye and filter that duplicates the penguin's retinal fluid. The orange dye has been used to produce orange-tinted sunglasses which provide improved vision in bright sunlight and on foggy days.

According to Donald DeYoung's book *Discovery of Design*, many welders now use orange-colored masks that are safer and more transparent than the old-style dark masks that made it difficult for them to see. There is also the hope that orange-tinted glasses may someday help patients suffering from visual loss due to cataracts or macular degeneration.

When engineers and designers look at nature to design new products and product improvements, they are looking at intelligent designs, not the products of chance and billions of years. The "sunglasses" worn by penguins were designed by their Creator!

Prayer: Father, I am grateful that You have given me not only eyes to see the world around me but spiritual eyes that can see the truths in Your Word! Amen.

Ref: "Penguin Eye – Sunglasses," *Discovery of Design,* D. DeYoung and D. Hobbs, pp. 112-113 (Master Books, Second Printing, 2012).

Your Potato Chip Bag Is Listening

Ecclesiastes 2:11
"Then I looked on all the works that my hands had wrought, and on the labour that I had laboured to do: and, behold, all was vanity and vexation of spirit, and there was no profit under the sun."

If you're concerned about maintaining your privacy in a world that is bent on snooping into every aspect of your life, you may not like what scientists are working on now. Researchers at MIT, Microsoft, and Adobe have developed a computer algorithm that can reconstruct an audio signal by analyzing the small vibrations of objects depicted in video. In other words, your potato chip bag is recording your words.

According to *ScienceNews*, the researchers were able to recover intelligible speech from the vibrations of a potato-chip bag that they filmed through soundproof glass from fifteen feet away. They were also able to extract audio signals from videos of aluminum foil and the surface of a glass of water. In fact, they could even recreate sounds that were filmed on the leaves of a potted plant.

Alexei Efros, an associate professor of computer science at the University of California, Berkeley, could hardly contain his excitement: "We're scientists," he said, "and sometimes we watch these movies, like James Bond, and we think, this is totally out of some Hollywood thriller. You know that the killer has admitted his guilt because there's surveillance footage of his potato chip bag vibrating."

Creation Moments can't help but be impressed by the ingenuity of scientists who develop surprising new technologies. After all, man's ingenuity points to the awesome genius of God – the One who created man in the first place!

Prayer: Lord, I pray that more scientists will develop an interest in You and not just Your creation. Amen.

Ref: "Extracting audio from visual information: Algorithm recovers speech from vibrations of a potato-chip bag filmed through soundproof glass," *ScienceDaily*, 8/4/14. Source: Massachusetts Institute of Technology.

Disposable Diapers for Robins

Luke 9:58
"And Jesus said unto him, Foxes have holes, and birds of the air have nests; but the Son of man hath not where to lay his head."

Parents, wouldn't it be nice if you didn't have to deal with dirty diapers during the early years of your child's life? And think of the money you would save! Too bad your little angel isn't more like a little robin.

Some birds, including the familiar robin, have it all figured out. You see, just seconds after a young robin has eaten, he eliminates waste into what can best be described as a white disposable diaper. All the parents have to do is pick it up and fly off with it, leaving the nest neat and clean.

This disposable diaper is called the fecal sac. It is made of thick, strong mucus that the parent can pick up and dispose of without puncturing it with its sharp beak. Robins will usually drop the sac twenty to fifty yards away before returning to the nest with another meal for its chicks. Once the young robin has matured enough to leave its nest, it no longer produces the fecal sac. Instead, its droppings are disposed of most often on your car's windshield.

Now, just imagine what a robin's nest would look like if their Creator hadn't come up with this ingenious plan. Without this disposable diaper, the nest would quickly become unsuitable for life. So the next time you find an abandoned robin's nest, take a peek inside. The only reason it's so clean is because God provided the robin with a disposable diaper service!

Prayer: Heavenly Father, the more I learn about Your creation – even the robin and its young – the more I am filled with praise for You! Amen.

Ref: "Disposable Diapers for Birds: The Scoop on Poop," Journey North.

Are You Superstitious?

1 Timothy 4:7
"But refuse profane and old wives' fables, and exercise thyself rather unto godliness."

Are you superstitious? A Harris poll conducted early in 2014 found that: 21 percent of Americans believe that knocking on wood prevents bad luck; 20 percent wouldn't dream of walking under a ladder; 14 percent believe it's unlucky to open an umbrella indoors; and 12 percent think the number 13 is unlucky. In fact, 14 percent of Americans said they believe Friday the 13th is an unlucky day.

Why are people superstitious? For the most part, superstition is caused by fear of bad luck and insecurity in the future, writes Benjamin Radford, *LiveScience* Bad Science columnist. But as you're about to hear, superstitions are not just foolish, they can be quite dangerous.

In May, a man fainted while driving through a tunnel near Portland, Oregon. This caused a head-on collision that sent four people to the hospital. The driver said he had lost consciousness while holding his breath. According to CarMax, the auto sales giant, "Holding your breath while driving though a tunnel is [America's] favorite driving superstition."

Christians, don't be superstitious. You have nothing to gain by being superstitious. Some people would even say that superstition is just another word for idolatry. As believers, we should not put our faith in objects like rabbits' feet or in manmade rituals like holding our breath in a tunnel. We should put our faith only in the one true God who is sovereign over all His creation.

> **Prayer: Heavenly Father, I place my trust in You and You alone. Though superstitions appear harmless on the surface, I know deep down that they do not bring glory and honor to You. Amen.**

Ref: Benjamin Radford, "Breath-Holding Superstition May Have Caused Car Crash," *LiveScience*, 6/5/14.

A Knotty Solution

Romans 5:6
"For when we were yet without strength, in due time Christ died for the ungodly."

"That's a knotty problem!" This figure of speech is used to describe a problem that's complex and difficult to solve – filled with knots, so to speak. Today, we're going to talk about a knotty *solution* that God has given to a fish that's been called the most disgusting creature in the sea.

The hagfish has a rather nauseating form of self-defense. As soon as a predator disturbs it, the little creature instantaneously releases slime into the water. This slime quickly clogs the predator's mouth and gills, causing it to beat a hasty retreat. The microfibrous slime absorbs water, creating a huge amount of the gooey stuff. In fact, a hagfish can fill a large home aquarium with slime within a matter of seconds.

But the slime isn't healthy for the hagfish either. What is he going to do to keep from clogging his own gills? His Creator has given this creature the ability to literally tie itself into an overhand knot by curling its own body around itself. The fish then manipulates the knot toward its other end, scraping the slime off its body.

The knotty solution that God has given to the hagfish can't begin to compare to the solution that God has given us to escape from the knotty problem of sin. We don't have the ability to rid ourselves of the sin that clings to us like the slime on a hagfish. That's why God removes the penalty and power of sin when we put our faith in Jesus.

Prayer: Heavenly Father, thank You for sending Your Son into the world as the one and only solution to the knotty problem of sin. Amen.

Ref: "Creatures of the Deep Sea: Atlantic Hagfish," Sea and Sky.

Seven Flagellar Motors in One

Psalm 141:1
"LORD, I cry unto thee: make haste unto me; give ear unto my voice, when I cry unto thee."

Try as they might, evolutionists are powerless to account for the microscopic rotary motor that propels bacteria through fluid just like a powerboat skims along the surface of a lake. But if the bacterium's flagellar motor troubles evolutionists, they must *really* be puzzled by the fast-moving MO-1 bacterium. After all, its hair-like flagellum isn't powered by just one motor. It uses seven motors all hooked up in parallel!

If the MO-1 bacterium were the size of a small speedboat, its proportional speed would be ten times the speed of sound. You'll need a microscope to actually see one. The bacterium measures only about 225 nanometres wide, so you'd need forty-four of them side by side to amount to the width of a single grain of talcum powder.

According to a team of researchers working in France and Japan, the flagellar apparatus of marine bacterium MO-1 is a tight bundle of seven flagellar filaments enveloped in a sheath. The motors are arranged in an intertwined hexagonal array similar to the thick and thin filaments of vertebrate skeletal muscles. There are also twenty-four fibrils in the sheath that are thought to counter-rotate between the flagella to minimize the friction of high-speed rotation.

From the Scriptures we see that God moves quickly when we call upon Him in prayer. God is never far from us, of course. But He is never nearer than when we draw near to Him in prayer.

Prayer: Heavenly Father, how often I have failed to call upon You when troubled by life's circumstances. Forgive me, Lord. Let me think of You as the first rather than the last resort! Amen.

Ref: "Architecture of a flagellar apparatus in the fast-swimming magnetotactic bacterium MO-1," *PNAS* (Proceedings of the National Academy of Sciences), Ruan et al, www.pnas.org/cgi/doi/10.1073/pnas.1215274109. "The Multiple flagellar Motor, evidence of evolution, or design?" 8/2/14, http://elshamah.heavenforum.org/t1850-multiple-flagellar-motors#3091.

Dinosaur-to-Bird Evolution Challenged

Psalm 1:6
"For the LORD knoweth the way of the righteous: but the way of the ungodly shall perish."

Most evolutionists today hold to the belief that modern-day birds evolved from dinosaurs. In fact, they now feel that large, carnivorous theropods like T. Rex rapidly shrank over a period of 50 million years until they evolved into our fine feathered friends.

But not all evolutionists agree with this belief. Some, in fact, claim that birds evolved into dinosaurs. A study published in *PNAS*, the Proceedings of the National Academy of Sciences, provides evidence that birds did not descend from ground-dwelling theropod dinosaurs.

According to Oregon State University zoology professor John Ruben, the research was well done and consistent with a string of studies in recent years that pose an increasing challenge to the birds-from-dinosaurs theory. The weight of the evidence, he added, is now suggesting that not only did birds *not* descend from dinosaurs, but that some species now believed to be dinosaurs may have descended from birds!

Professor Ruben's most revealing comment, however, was that the old dinosaurs-to-birds theories, instead of carefully interpreting the data, had public appeal, and that "many people saw what they wanted to see."

How true for scientists on *both* sides of this issue! When scientists start with the assumption that evolution is true, then every bit of evidence they uncover will support evolution. They don't even consider a third option – that God made the birds on the fifth day of Creation Week and the land-dwelling dinosaurs on the sixth day.

Prayer: Heavenly Father, while people say that the Bible is not a science textbook, it does provide information that scientists would be wise to take into consideration. Open their eyes, I pray. Amen.

Ref: "Bird-from-dinosaur theory of evolution challenged: Was it the other way around?" *ScienceDaily*, Source: Oregon State University, 2/10/10.

The Fish That Predicts Earthquakes

Revelation 16:18
"And there were voices, and thunders, and lightnings; and there was a great earthquake, such as was not since men were upon the earth, so mighty an earthquake, and so great."

Behold the giant oarfish – a creature of which legends are made. With its snake-like body that can grow up to fifty-six feet in length, the longest bony fish in existence has inspired countless legends about sea serpents.

Strangely enough, one of these legends just happens to be true. Japanese folklore relates that oarfish were the messengers from the sea god's palace that were able to predict a tremendous catch of fish or a disastrous earthquake. While we can't say they are good luck for fishermen, their earthquake-forecasting ability is a verifiable fact.

Numerous sources cited at the CreationWiki website relate that unusual numbers of dead oarfish have been observed to wash up on shore shortly before an earthquake. In Japan, for instance, whenever an earthquake followed by a tsunami is about to hit, several oarfish wash up on shore dead, indicating the disaster is imminent.

Many of you listening to me now will remember the devastating earthquake that hit Japan in 2011 and the tsunami that caused the catastrophic Fukushima nuclear reactor meltdown. You probably don't know, however, that many oarfish washed up on shores just before the earthquake struck. Scientists believe that this happens because the oarfish's large body makes it vulnerable to underwater shock waves or because poison gases are released during seismic activity.

Whatever the cause, the oarfish is not a sea serpent we should fear. The oarfish is one of God's creatures we should pay attention to!

Prayer: Father, You have made everything for a purpose. While that purpose is often not clear to us, help us to trust You no matter what the circumstances. Amen.

Ref: http://creationwiki.org/Oarfish.

Ancient Romans and Nanotechnology

Hebrews 2:6b-7a
"What is man, that thou art mindful of him? or the son of man, that thou visitest him? Thou madest him a little lower than the angels;"

One of the most interesting artifacts at the British Museum is the Lycurgus Cup, a beautiful goblet made by the Romans sixteen hundred years ago. The cup depicts the story of King Lycurgus, entangled in grapevines for his treachery against Dionysus, the god of wine in Greek mythology.

But it's not the design that makes the cup of such great interest to scientists. When viewing the cup in normal lighting, the glass is jade green. But when the cup is lit from behind, the glass changes to a bright ruby red. What is this ancient sorcery that gives the cup its color-changing properties? It's the precise amount of silver and gold nanoparticles dispersed throughout the glass material.

New research has discovered that the cup even changes colors when liquid is poured into it. Scientists replicated the makeup of the cup and then filled it with various liquids. Their results suggest the cup might have displayed many different colors depending on what sort of beverage was poured into it.

As one science website puts it, "...researchers are just now, all these years later, learning about such color changing properties of materials with embedded nanoparticles. The hope is that these properties can be exploited to perform chemistry or medical tests cheaply and quickly by displaying different colors under different conditions."

Creationists need not wonder how the ancient Romans knew about nanotechnology. However far back you go in history, the ingenuity of man – including "ancient man" – points to the even greater ingenuity of his Creator!

Prayer: Father, thank You for ancient artifacts revealing that You gave early mankind more abilities than evolutionary scientists ever expected. Amen.

Ref: Bob Yirka, "Goblet tricks suggests ancient Romans were first to use nanotechnology," Phys.org, 8/27/13.

The Drake Equation

Ephesians 6:10-11
"Finally, my brethren, be strong in the Lord, and in the power of his might. Put on the whole armour of God, that ye may be able to stand against the wiles of the devil."

How many times have you seen headlines proclaiming: "Sixty billion planets in the Milky Way could support life" or "One hundred billion planets in our galaxy may harbor complex life"? These numbers change with each new announcement. Where did those numbers come from in the first place?

It all started when astronomer Frank Drake, a firm believer in extraterrestrial life, hosted the Search for Extraterrestrial Intelligence Institute's very first meeting back in 1961. He wanted to inspire the other scientists in attendance, so he developed an equation which has been described as a "probabilistic argument used to estimate the number of active, communicative extraterrestrial civilizations in the Milky Way galaxy."

The Drake Equation, as it has come to be called, sounds mathematical and scientific, but when you examine the equation, you will see that it is based on conjecture after conjecture, guess after guess. In fact, most scientists today totally reject it, calling it a guesstimate or meaningless. Nevertheless, we keep seeing headlines about the fifty or sixty or one hundred billion planets out there which are capable – or even likely – of having life. After all, if life evolved on Earth, surely it evolved on many planets which may be "capable" of supporting life.

But as scientist Donald DeYoung points out in his book *Astronomy and the Bible*, the first variable in the Drake Equation actually turns out to be zero. So when you multiply all the variables together, you end up with not billions but a big fat zero!

Prayer: Heavenly Father, I pray that You will help me develop critical thinking skills and spiritual wisdom so that I will be fully equipped to detect error when false teachers seek to deceive me. Amen.

Ref: D. DeYoung, *Astronomy and the Bible,* pp. 117-118 (BMH Books, 2010). Wikipedia entry on "Drake equation."

Cloudy with a Chance of Crayfish

Acts 1:11b
"Ye men of Galilee, why stand ye gazing up into heaven? this same Jesus, which is taken up from you into heaven, shall so come in like manner as ye have seen him go into heaven."

Though we are aware of no confirmed reports of meatballs falling from the sky during a rainstorm, there have been hundreds of reports of strange things falling from the clouds during strong storms. And sometimes, this even includes living animals.

According to the Weather Channel, one such instance happened in 2013 when live crabs fell on the community of Lynn Haven, Florida. Another storm of living creatures happened on June 28, 1957, near Thomasville, Alabama. This was when thousands of live fish, frogs, and crayfish fell from the sky during a heavy rain. It was thought that an F2 tornado fifteen miles to the south was responsible for this strange phenomenon.

Creationist Donald DeYoung writes about a storm in 1984 that dumped live six-inch flounder in a London suburb. Rains of flightless animals have been reported throughout history. In the first century AD, Roman naturalist Pliny the Elder documented storms of frogs and fishes. Indeed, the most common type of creature to fall from the sky is fish. It is thought that they get picked up by waterspouts and are then dropped over land, sometimes miles away.

While there are hundreds of stories like these from all over the world, the most unusual living creatures ever to appear out of the sky were horses of fire just before Elijah the prophet was taken up by a whirlwind into the heavens. But even that extraordinary historic event will pale in comparison to the return of Jesus Christ in the clouds.

Prayer: Heavenly Father, just as Your Son ascended to heaven, I know that He will return to the Earth in the same way. Lord, I eagerly await that most blessed of days. Amen.

Ref: "World's Strangest Weather Phenomena," The Weather Channel, 4/7/14. "Has it ever rained frogs or fish?" D. DeYoung, *101 Questions and Answers About Weather and the Bible*, p. 84 (Baker Books, 2014). Wikipedia entry on "Raining animals."

"Was I Going to Be Arrested?"

Acts 1:8
"But ye shall receive power, after that the Holy Ghost is come upon you: and ye shall be witnesses unto me both in Jerusalem, and in all Judaea, and in Samaria, and unto the uttermost part of the earth."

One of our listeners – pastor and creationist Phil Spry – told us a remarkable story that I'm going to share with you in his own words.

Shortly after the old USSR dissolved I was in Moscow teaching Creation Science seminars in Russian high schools. At one school the missionary who set up the meeting of about seven students had to go to the school office to get a screen for my overhead projector. He came back with the school principal, who wanted to know what we were doing in that 800-seat auditorium with just seven kids. My translator told him, "This American has come here to tell the students how God created the world." He got very agitated and told us to stop. He quickly exited the room through a side door. I thought I was going to be arrested.

The principal went down the hall to the first classroom, opened the door and said to the teacher, "Bring all of your students and come with me." He then went to every classroom in the building and did the same thing. It took about thirty minutes until every student, teacher, administrator, custodian, counselor and cook were in the auditorium. He then came to me and in animated Russian spoke to my translator: "For seventy years our country has lost the knowledge of God. Please; tell my students how God created the world." Then he gave me the rest of the day with about a thousand people.

Creationists, do not lose heart! That day I shared the message of Genesis chapter one, but more importantly, of John chapter three!

Prayer: Father, thank You for opening doors for this man. I pray You will also open doors for me so I can share biblical creation with many. Amen.

Ref: Correspondence with creationist, church planter, author and pastor Phil Spry. Used with permission.

Antzilla!

Matthew 7:12
"Therefore all things whatsoever ye would that men should do to you, do ye even so to them: for this is the law and the prophets."

Welcome to Australia! This delightful continent is home to many of the world's most poisonous snakes and spiders, not to mention deadly sea creatures like the box jellyfish, salt water crocodiles, and everybody's favorite – the great white shark. Today, however, we're going to talk about another creature you don't want to meet on your next trip down under. It's called the jumper ant because it can jump as high as two inches into the air. But most people call it the bulldog ant, and with good reason.

These aggressive ants got their name from their powerful sting, strong grip, and savage biting behavior. Bulldog ants are fast and they're big, growing up to 1.6 inches in length. This makes them among the largest ants in the world. Their eyes are big, too, so they don't miss a thing. Their long mandibles are menacing, especially with their rows of long teeth.

But they're even *more* dangerous at their posterior end, where their venomous stinger is located. The ant's aggressive behavior is so famous, it was even immortalized in one of philosopher Arthur Schopenhauer's major works. "If it is cut in two," he wrote, "a battle begins between the head and the tail. The head seizes the tail in its teeth, and the tail defends itself bravely by stinging the head."

While aggressive behavior is the rule in a world that has been corrupted by sin, Christians exhibit a different kind of behavior – looking out for the benefit of others.

Prayer: Heavenly Father, I pray that You will constantly remind me – especially when I am hurt by others – that I am not to return evil for evil. I am to return evil with forgiveness and love. Amen.

Ref: "Bulldog Ants," *National Geographic*, May 2016. Anindita Roy, "Bulldog ant," Animal Spot. *The World as Will and Representation*, A. Schopenhauer.

Mercury Drives Zebra Finches Crazy

1 Samuel 21:13
"And he changed his behaviour before them, and feigned himself mad in their hands, and scrabbled on the doors of the gate, and let his spittle fall down upon his beard."

A few days ago, a light bulb on my ceiling fan exploded, showering my room with thousands of large, small, and tiny pieces of glass. I breathed a sigh of relief because I hadn't been hit by any flying glass. But there's another reason I was grateful.

At a recent Animal Behavior Society meeting at Princeton, John Swaddle of the College of William and Mary told how he and his colleagues had exposed zebra finches to an amount of mercury simulating a contaminated environment. The poisoned birds were bolder and hyperactive but spent less time feeding. Swaddle said that their hyperactivity would put them on the dementia side of mercury poisoning.

He said this because small doses of mercury are known to cause dementia in humans. And this is why I was grateful that the light bulb which had exploded near me was a regular incandescent bulb. If it had been a new compact fluorescent light bulb, I would have been exposed to poisonous mercury vapor. In the United States, the Environmental Protection Agency has a ten-step advisory on how to handle broken CFL bulbs. All I had to do was sweep up the broken glass.

Since Alzheimer's disease and dementia are rising, one has to wonder why the United States would ban incandescent bulbs and insist that they be replaced with light bulbs containing mercury, a substance known to cause depression and dementia.

While Creation Moments supports good stewardship of the environment, we do not support actions that put our health at risk. After all, we were made in the image of God.

Prayer: Father, I thank You for new technologies that benefit mankind while, at the same time, help us to be good stewards of the world You created. Amen.

Ref: S. Zielinski, "Zebra finches go mad with mercury, and other animal updates," *ScienceNews*, 8/17/14. United States Environmental Protection Agency, "Cleaning Up a Broken CFL."

What If Noah's Ark Is Found?

2 Timothy 3:12
"Yea, and all that will live godly in Christ Jesus shall suffer persecution."

What if explorers actually found Noah's Ark? Would it instantly change unbelievers into followers of Jesus? Would scientists suddenly drop their belief in an old Earth and millions of years of evolution and start believing the Bible?

No, they would not. We've already seen how evolutionary-minded universities treat scientists who discover soft tissue in dinosaur remains, showing that dinosaurs couldn't possibly be millions of years old. They fire such scientists. Many other instances of the persecution of Darwin-doubting scientists can be found in Dr. Jerry Bergman's excellent book, *Slaughter of the Dissidents*. In this book he shares story after story of scientists who have lost their jobs, who have been denied tenure, and whose careers have been abruptly ended.

Dr. Bergman's book is dedicated "to those who have paid a high price for the discrimination they have suffered while taking a stand for freedom." Indeed, Bergman relates how creationists are often demoted, transferred to other departments, and ridiculed, while some even receive death threats.

Discrimination and persecution of Christians are on the rise all over the world. In some countries, Christians aren't just losing their jobs, they're losing their lives. We must pray for them, and we must not be caught by surprise when persecution comes to our door. When our faith is tested in the crucible of fire, we must pray for God's grace, knowing that He will have the victory no matter what happens to us. To God be the glory!

Prayer: Heavenly Father, You promise in the Bible that Your people will suffer persecution in this world. I do not ask You to help me escape the persecution. Rather, I ask You to give me the grace to endure persecution without wavering in my faith. Amen.

Ref: J. Bergman, *Slaughter of the Dissidents*, Volume 1, 2nd Edition (Leafcutter Press, 2012).

Trash-Talking Bats

Psalm 71:1
"In thee, O LORD, do I put my trust: let me never be put to confusion."

Some sports figures have practically made an art of trash-talking. Perhaps the most famous of them all is Muhammad Ali. His taunts-filled trash-talk – both before and during the bouts – worked so well, it helped him become heavyweight boxing champion.

Until recently, no one knew that some bats do the very same thing in order to distract other bats from gobbling up bugs they want for themselves. Using high-pitched echolocation signals at just the right moment, the Mexican free-tailed bat can block a competitor's chance to get a meal.

As reported in *Science* magazine, a biology postdoctoral student at the University of Maryland discovered this behavior by accident. When he played a recording of a bat's echolocation signal right as a bat was about to catch an insect, the bat was up to 85.9 percent less likely to catch its prey.

A biologist at the University of Toronto who studies bat echolocation but who wasn't involved in the new study was quick to note that the bats' competition for the same food source provided "the perfect environment" for such a behavior to evolve.

It goes without saying that he reached such a conclusion without any evidence whatsoever. The minds of far too many biologists are so permeated by evolutionary thinking that they are unable to even consider the possibility that bats were designed by a Creator who gave each creature the skills it needs to thrive.

Prayer: *Heavenly Father, someone once said that You must love bats because You created so many different kinds! Your greatest love is reserved for those You adopt into Your family by regeneration and new birth in Your Son. Amen.*

Ref: "Bats Make Calls to Jam Rivals' Sonar – First Time Ever Found," National Geographic online, 11/6/14. A.J. Corcoran and W.E. Conner, "Bats jamming bats: Food competition through sonar inference," *Science*. Vol. 346, November 7, 2014, p. 745.

Evolution in Reverse

Romans 3:20
"Therefore by the deeds of the law there shall no flesh be justified in his sight: for by the law is the knowledge of sin."

Some time ago, we told you about the nineteenth-century paleontologist Louis Dollo, who proposed a law that has become a cornerstone of evolutionary belief. Dollo's law says that a structure or organ lost during the course of evolution would not reappear in that organism. In other words, evolution never shifts into reverse.

But even a recent issue of *Smithsonian Magazine* points out that Dollo's law has been broken again and again. For instance, it mentions a tree frog from South America that lost its lower teeth only to re-evolve them after 200 million years.

Apparently, Dollo's law has now been broken yet again. According to a recent study of the wrists of modern birds, a bone lost from dinosaurs for tens of millions of years reappeared when dinosaurs evolved into birds and took flight.

But wait! According to Dollo's law, evolution *never* goes backward. Structures which have disappeared should not return. Ever faithful to Darwinism, the magazine is left to conclude: "Perhaps the very bone you are sitting on, your coccyx, is ready to re-evolve a tail at some future moment when humans might need it again to hang from trees."

Dollo's law should be true if you're an evolutionist … but it isn't. However, the law established by our Creator is true, and it condemns each and every one of us because we have broken that law. Christians can thank God for that law because it drives us to seek salvation in Christ.

Prayer: Lord Jesus, thank You for taking upon Yourself the punishment I deserve for breaking Your commandments. Because of Your sinless life, Your death on the cross and Your resurrection from the dead, I can look forward to eternal life with You! Amen.

Ref: "The Wrists of Birds Reveal Evolution Undoing Itself," Smithsonian.com, February 2015.

On Bended Knees

Romans 14:11
"For it is written, As I live, saith the Lord, every knee shall bow to me, and every tongue shall confess to God."

While attending school in England, I was taught that my knee is a hinge joint – a rather simple design, much like the hinge of a door. But I now know that the knee is actually a masterpiece of design that could only have come from the hand of a supremely intelligent Designer and Creator.

Dr. Stuart Burgess, the aerospace engineer responsible for the solar array deployment system on the European Space Agency's earth observation satellite system, wrote a whole chapter about the human knee in his book *Hallmarks of Design*. He refers to the knee as a sophisticated "four-bar system" in which two bones and two ligaments work together in perfect harmony.

"When a mechanical engineer looks at the anatomy of the human leg," Burgess writes, "the four-bar mechanism in the knee joint stands out as one of the most important and impressive mechanisms." He observes that while the ball-and-socket joint is explained to students in England, the four-bar mechanism of the knee is rarely mentioned in biology textbooks. Burgess believes that teachers are reluctant to talk about it because such a sophisticated mechanism bears the hallmarks of design.

Someday every knee shall bow before the Creator of the universe ... and this includes even the most obstinate evolutionists. Creation Moments is here to help you share the truth of biblical creation with unbelievers so that when they do bend their knees, they will do so not out of force but in worship.

Prayer: Heavenly Father, I pray for opportunities to share the truth of biblical creation. Make me bold in standing up for Your truth in the midst of an unbelieving culture. Amen.

Ref: S. Burgess, *Hallmarks of Design*, p. 14 (Day One Publications, 2002).

An Open-and-Shut Case for Creation

Genesis 6:17
"And, behold, I, even I, do bring a flood of waters upon the earth, to destroy all flesh, wherein is the breath of life, from under heaven; and every thing that is in the earth shall die."

Evolutionists hate the very idea of the worldwide flood because they know that such a catastrophe would cause the Earth to appear to be much older than it truly is. And even they know that without those long ages of time, evolution becomes an impossibility.

Surprisingly, dead clams – like many other sea creatures – have much to tell us about the flood of Noah's time. Clams are known as bivalve mollusks. They have two valves that are hinged together, and they can close their shells very quickly to protect themselves from predators. When a clam dies of natural causes, the clam releases its grip on its shell, and it opens within a matter of hours. But if a buried clam shell is found shut tight, it shows that the clam was buried suddenly.

We mention this because fossil clam graveyards are found all over the world containing millions of clams with their shells shut tight. This speaks to us of a catastrophe that buried millions of clams very quickly. In other words, the worldwide flood of Noah's time. Since these fossil clam beds are found all over the world – and are often high above their natural habitat – it can only mean that the biblical worldwide flood is a fact of history.

Now, if the Bible is right about the worldwide flood, the Earth is much younger than it appears to be, so there simply isn't enough time for evolution to take place. Graveyards of closed clams have given us an open-and-shut case for biblical creation!

Prayer: Heavenly Father, thank You for providing abundant evidence that the Bible is true. I pray that You will remind me of these things when I am talking to Bible skeptics. Amen.

Ref: "Clams," Evidence of Design online, 9/11/09.

Did We Once Walk on All Fours?

> ***Genesis 1:24***
> *"And God said, Let the earth bring forth the living creature after his kind, cattle, and creeping thing, and beast of the earth after his kind: and it was so."*

Evolutionists tell us that we used to walk on all fours, just like the apes from which we evolved. While this is fine for the *Planet of the Apes* science-fiction films, creationists are adamant that such a theory should not appear in school textbooks.

Family physician and author Dr. Robert Peprah-Gyamfi would agree. In his book *Seeing God Through the Human Body,* he recalls how he was taught in school that humans once walked on all fours. Even before he became a Christian, he didn't believe it for one second. But now, as a Bible-believing creationist, he regards that notion as absolutely ridiculous.

In the book's chapter on the human skeleton, he points out that God "built the human skeletal system as a kind of scaffold that would not only support and protect delicate organs of the body, but also enable us to move from place to place."

He went on to write that evolution is "a preposterous theory that stems from nothing other than unbelief propagated by people who cannot or are not prepared to believe there is a powerful force in the universe, God Almighty, capable of doing anything." He sums it all up with these words: "The whole world may take me for a naïve, crazy person running away from the bare facts. As far as I'm concerned, however, the notion that we once walked on all fours is absolute rubbish."

Don't worry, doctor. There are millions of people who believe exactly as you do!

Prayer: Heavenly Father, I know that when Adam and Eve were expelled from the Garden of Eden, each of them walked out on two feet, not four! Amen.

Ref: Dr. Robert Peprah-Gyamfi, *Seeing God Through the Human Body*, pp. 127-132 (HighWay, 2009).

Our Multi-Directional Genetic Code

Romans 11:33
"O the depth of the riches both of the wisdom and knowledge of God! how unsearchable are his judgments, and his ways past finding out!"

Have you ever heard the palindrome "Madam, I'm Adam"? With a palindrome, the letters are exactly the same whether you read the sentence from left to right or from right to left.

But did you know your genetic code is a different kind of palindrome that conveys one series of messages when it is read from front to back and a different series of messages when it is read from back to front? Other meaningful information can even be found in overlapping portions of the genome! Think about your genetic code as if it were a novel. It would tell one story when read from front to back, a *different* story from back to front, and many other stories as you read overlapping messages of the code!

Evolutionists find this very troubling. After all, their whole theory depends on random mutations resulting in improvements. But as scientists at the FMS Foundation point out, this multi-directional coding makes it impossible for a random "typo" to benefit one code without disrupting the others. This fundamental problem, they said, has been overlooked until now. "The bottom line," they conclude, "is that overlapping codes cannot arise by any natural process – they must be very carefully designed."

Yes, evolutionists must come to grips with the fact that the genetic code was designed. Chance mutations would only destroy the code. While evolution can't even begin to account for the genetic code found within every cell of your body, it was an easy task for our Creator!

Prayer: Oh Lord, thank You for revealing to us the incredible complexity of the genetic code. I pray that You will use this information to utterly destroy the foolishness of evolution. Amen.

Ref: Chris Rupe, Dr. John Sanford and Chloe Pappano, "Scientists Discover Astounding DNA Complexity," FMS Foundation, 2014.

The Brilliance of Butterfly Wings

1 John 4:6
"We are of God: he that knoweth God heareth us; he that is not of God heareth not us. Hereby know we the spirit of truth, and the spirit of error."

What do a butterfly's shimmering wings, a fish's opalescent scales, and a peacock's brilliant feathers have in common? None of the colors come from pigments. As an article at the Gizmodo website rightly points out, "All of their beautifully iridescent colors are produced by the physical interaction of light with sophisticated nanoscale architecture that we are only just beginning to understand."

That is a true statement, a scientific statement describing why such striking colors come from wings that do not use pigments. Unfortunately, the writer quickly turns from science to science fiction when she states that the first color-building life forms probably date back to the Cambrian explosion over 500 million years ago – a time that, quote, "was a period of rapid evolutionary advancement, during which eyes evolved from simple light sensors to neurologically sophisticated organs that could detect color, shade and contrast."

Where is the evidence for such a statement? And why can't science writers stick to science? Once again, we see how science writers – and the publications they work for – seamlessly blend science and science fiction into the same sentence. Evolutionists' brains are wired this way because of their commitment to the philosophy of evolution. So, too, should Christians be so committed to the authority of the Word of God that we can't help but view the vibrant colors of butterfly wings as painted there by our magnificent Creator!

Prayer: Heavenly Father, help us distinguish between truth and error in everything we read, hear and see. We rely upon Your Spirit of Truth to guide us in these dark, deceptive and troubling times. Amen.

Ref: M. Stone, "How Nanoscale Optics Create Nature's Most Dazzling Colors," *Gizmodo*, 2/19/15.

Come Out, Come Out, Wherever You Are!

Psalm 119:114
"Thou art my hiding place and my shield: I hope in thy word."

It has long been known that electric eels are able to immobilize prey with a jolt of electricity. But new research, reported in the journal *Science*, has shown that eels also use their electric organs to make fish come out of hiding. A researcher from Vanderbilt University in Nashville, Tennessee, found that the electric discharges from eels caused the muscles of their prey to twitch, making them easier to capture by revealing their whereabouts.

Dr. Kenneth Catania, who led the study, set up small aquatic arenas to test the eels' hunting abilities – putting an eel and a fish into the same tank. As he discovered, the eels' electric pulses directly activated the nerves that controlled their prey's muscles. "You and I couldn't activate every muscle in our bodies at once," he said, "but the eels can do that [remotely] in their prey."

Dr. Catania noticed that hungry eels would emit pairs of pulses when their potential meal was out of view. These "doublets", as they are called, generate very rapid and strong muscle contraction, causing the fish to jump and bringing them out of hiding.

Just like the unfortunate fish in the tank, we have powerful enemies who seek our destruction. But we know we also have an all-powerful God who protects and preserves us. In Psalm 143:9, the writer cries out: "Deliver me, O LORD, from mine enemies: I flee unto thee to hide me." While our enemies may be many and strong, our God is stronger still.

Prayer: Heavenly Father, thank You for being a very present help in times of trouble. When I am under attack, I know I can turn to You. Amen.

Ref: Victoria Gill, "Electric eels 'remotely control their prey'," *BBC News Science & Environment*, 12/4/14.

Evolutionists Are Simple Minded

Psalm 9:1
"I will praise thee, O LORD, with my whole heart; I will shew forth all thy marvellous works."

Evolutionists are simple minded. It's a fact! Charles Darwin thought that the cell was virtually as simple as a blob of gelatin with a nucleus inside, and evolutionists have been simple minded ever since.

To check this out, Creation Moments asked an evolutionist which came first: the heart, the blood or the blood vessels? After all, if the heart evolved first, what good would it be without the other two? If the blood evolved first ... well, you get the idea. The evolutionist pretty much avoided our question and said that the heart could easily have come about through natural processes. After all, he said, the heart is "a simple pump."

To put it simply, evolutionists think they can bolster their argument by claiming that the incredibly complex things God created actually have quite simple beginnings. Random collisions between molecules – given enough time – will certainly be able to create simple things, they say. And those simple things, they continue to conjecture, will gradually become more and more complex things, given enough time. Case closed.

How different this is from reality! What scientists once called the "simple cell" is now known to be a city in miniature, complete with manufacturing facilities, transportation systems, power plants and a library of code that enables the cell to be self-replicating. And it's all packed inside that tiny so-called "simple" cell!

For those whose minds are not held captive by atheistic naturalism, an incredibly complex universe filled with amazingly complex creatures requires an unfathomably complex God.

Prayer: Lord Jesus, I pray that You will open the eyes of unbelievers so they will see the world around them as Your creation. Once they see this, I pray they will understand that You entered Your creation to die for their sins. Amen.

Ref: Steve Schwartz, "Evolutionists Are Simple Minded," Creation Moments, 4/26/11.

Light Bends!

Genesis 1:16
"And God made two great lights; the greater light to rule the day, and the lesser light to rule the night: he made the stars also."

Perhaps you'll remember that it was Arthur Eddington who measured the bending of starlight near the sun during an eclipse in 1919. His work validated details of Einstein's General Theory of Relativity and established Eddington as one of the greatest astronomers of all time.

But you may not have ever heard that this great scientist was also a man of great faith. His whole life was influenced by his reverence for God as Creator. In fact, he wrote that the details of Creation were worthy of being measured, explored and appreciated. In his book *Science and the Unseen World*, Eddington declared that the world's meaning was not discoverable from science but must be sought through understanding spiritual realities. How many scientists today would come out and say something like that?

As author and creationist Don DeYoung points out in his book *Pioneer Explorers of Intelligent Design,* Eddington avoided offering proofs for God's existence but said that the most flawless proof for the existence of God is no substitute for a personal relationship with Him. Such a relationship would take even the most convincing arguments against God's existence and turn them harmlessly aside.

Eddington was absolutely correct. A vital relationship with our Creator, Jesus Christ, will safeguard us from even the most persuasive claims of atheists. Not only will we know for certain that God exists, we will know from personal experience the peace that comes from having an intimate relationship with our Creator and Savior.

Prayer: Heavenly Father, I pray for evolutionists so that when they turn to You in faith and are born again, their changed lives will have an impact on many others. Amen.

Ref: Don DeYoung, *Pioneer Explorers of Intelligent Design*, pp. 2-3 (BMH Books, 2006).

Prima Ballerina of the Sea

Genesis 1:20a
"And God said, Let the waters bring forth abundantly the moving creature that hath life..."

The long and skinny shrimpfish must surely be one of the strangest sea creatures to come from the hand of a Creator who never ceases to amaze us with His wonders. Of course, evolutionists claim that the shrimpfish is a product of millions of years of evolution. Let's take a close look at this remarkable creature and see for ourselves which view holds water.

Shrimpfish spend their lives in the ocean with their mouths pointed straight down and their tails above them in a vertical stance. Not only do they feed like this. They swim like this, too.

Now if you were to look at a cross-section of most fish, you'd see an oval or a circle. But a cross section of the shrimpfish looks very much like a slice of an airplane wing. Its wider leading edge tapers off to a thinner edge toward the belly. It's an aerodynamic shape that makes it easy for the fish to cut through the water. But let me add that it works only when the fish swims in its natural vertical position rather than head-first like other fish.

Evolutionist Frank Fish – yes, that's his name – claims that the usual fin found on the backs of fish actually moved to join the shrimpfish tail, making it easier for the shrimpfish to turn easily. "They do full pirouettes on their heads," he added. "I'd like to see a prima ballerina do that."

With its many design features, the shrimpfish provides big evidence for creation, not mindless evolution!

Prayer: Lord Jesus, I pray that You will open the eyes of unbelievers so that they will see Your hand in creation as well as Your nail-pierced hands on the cross. Amen.

Ref: Susan Milius, "That's how shrimpfish roll," *ScienceNews*, 2/7/15, p. 4.

Confrontation with a Radical Atheist

2 Peter 3:12
"Looking for and hasting unto the coming of the day of God, wherein the heavens being on fire shall be dissolved, and the elements shall melt with fervent heat?"

Creation Moments and friends of our ministry spend countless hours sharing their faith and the truth of biblical creation on social networking websites. One such creationist, Allen Dunckley, Th.D., recently ran into a person he describes as a radical atheist. The atheist laid down a challenge with these words: "Demonstrate a causal link to why your god needs to exist or does exist."

Dr. Dunckley was quick to respond with a question of his own: "Can you explain to us how is it that the positively charged protons in the nucleus of an atom don't repel each other since it is a fact that like charges do so? By the laws of physics, the atomic nucleus should not exist, yet it does."

He continued: "This fact is consistent with Colossians 1:17 that says, 'by Him (Jesus Christ) all things consist.' 'Consist' is from a Greek word which means 'held together.' So here is demonstrable empirical fact that God exists and that His existence is what holds the very fabric of matter together. So the existence of the material universe is dependent on the existence of the Creator Himself. All that needs to be done to destroy the universe is for God to literally 'let go' of matter and – poof – gone just like the Bible says in 2 Peter 3:12."

As you can see, Christians are using social networking websites to share their faith with people, some of whom live on the other side of the world. Does God want you to join them in declaring His truth unto the ends of the earth?

Prayer: Heavenly Father, I thank You for the many Christians who are laboring in a mission field that is unique to our time in history. In Jesus' Name. Amen.

Ref: From Allen J. Dunckley's post in the "Creationists Speak Out" page on Facebook, 12/18/14. Used with permission.

Master of DisgEYES!

1 Kings 22:30
"And the king of Israel said unto Jehoshaphat, I will disguise myself, and enter into the battle; but put thou on thy robes. And the king of Israel disguised himself, and went into the battle."

When the South American false-eyed frog is approached by a predator, it behaves in a way that evolution is powerless to explain. It turns its back on the predator.

So how is this a good strategy? In the time it takes to spin around, the frog's back is changing color to look like a pair of large, menacing eyes, complete with black pupils surrounded by blue irises. And it accomplishes this in only one second.

To complete its disguise, the frog raises its rear-end up high and tucks its hind legs underneath his fake eyes so the legs now look like a large mouth. Above the fake mouth appears what looks like a nose. The frog also lifts two toes on each back foot and curls them out so they look like claws. By moving his legs, the frog's fake mouth and claws appear to be in motion.

If this trickery doesn't work, the false-eyed frog releases a smelly, sticky substance from glands located inside of the gigantic fake eyes. This final defense tactic is usually enough to scare away all hungry predators.

It should be obvious to anyone with common sense that evolution can't account for this remarkable defense mechanism. After all, if the frog turned around *before* evolution had given it the ability to transform its back into a pair of menacing eyes, it would have died out long ago. Only evolutionists with blinded eyes can possibly believe that the false-eyed frog is a product of evolution.

Prayer: Oh Lord, thank You for creating a tiny frog that shows how foolish it is to attribute its amazing defense mechanism to random mutations and natural selection over millions of years. Amen.

Ref: "Wild Facts #32: Two-Faced False-Eyed Frog," Wild Facts. "The False-Eyed Frog," Alpha Omega Institute, 9/28/11.

The Infinite Monkey Theorem

Isaiah 35:5
"Then the eyes of the blind shall be opened, and the ears of the deaf shall be unstopped."

If an infinite number of monkeys sat down at an infinite number of typewriters, eventually one of these monkeys would type Shakespeare's play *Hamlet*. At least that's what French mathematician Émile Borel proposed in his infinite monkey theorem – an idea that scientists use to assert that given enough time, even evolution can happen.

Now, if anyone ever tosses the infinite monkey theorem in your direction, point out that a monkey has a one in twenty-six chance of typing the first letter correctly. Getting the second letter correct sees the probability rise to 1 in 676. After just the first 15 letters of *Hamlet*, the chance of getting the next letter right is one in thousands of trillions!

In 2003, the infinite monkey theorem was put to the test at the University of Plymouth in the United Kingdom. Researchers put six crested macaques in front of a computer and waited for them to start typing. Occasionally, in between urinating on the keyboard and smashing it with a rock, the monkeys typed out the letter "S" several times before tiring of the experiment.

Despite the sheer absurdity of the infinite monkey theorem, it is still being used to show that complexity can arise from random mutations. In fact, as recently as 1986, evolutionary evangelist Richard Dawkins brought it up in his book *The Blind Watchmaker*.

It never ceases to amaze us what evolutionists will believe so they can continue in their unbelief. Clearly, the infinite monkey theorem has done nothing but make monkeys out of those who still believe it!

Prayer: Heavenly Father, from the probability standpoint alone, evolution is absolutely impossible. I pray that You will make this very clear to those who are now stumbling about in the darkness of evolution. Amen.

Ref: R. Arp, "Infinite Monkey Theorem," *1001 Ideas That Changed the Way We Think*, p. 643 (Atria Books, 2013).

First Hero in the Bible's Hall of Faith?

Hebrews 11:6
"But without faith it is impossible to please him: for he that cometh to God must believe that he is, and that he is a rewarder of them that diligently seek him."

See if you can answer the following question from Creation Moments Board Director Mark Cadwallader. The eleventh chapter of Hebrews includes a list of heroes of the faith. Each verse that introduces a faith hero begins with the words "by faith" or "through faith." Can you name the very first hero mentioned in Hebrews 11? Also, can you identify why he or she was said to exercise faith?

Most people, of course, say the name of an Old Testament character. Many think it was Abel, who offered unto God a more excellent sacrifice than Cain. But the right answer is – wait for it – YOU! As a believer in biblical creation, YOU are at the head of the list!

Check out Hebrews 11:3 where it says, "Through faith WE understand that the worlds were framed by the word of God, so that things which are seen were not made of things which do appear."

That's right – believers are listed first in the Hebrews 11 "Faith Hall of Fame"! And we are listed there because we understand that God created the worlds through Jesus, the Word. The first faith hero in this amazing chapter are believers down through the ages who believe what God said about His act of Creation.

This means that Christian creationists have a responsibility to exercise the faith we have been given to share the reasons for our faith with others. What have you been doing lately to spread the truth of the gospel and biblical creation?

Prayer: Oh, Lord, make me bold in sharing my faith with others. Amen.

Ref: Mark Cadwallader, "Why Creation Evangelism Is Vital for Our Kids and Grandkids!" Creation Moments online.

Why You Were Born and Not Hatched

Job 10:18a
"Wherefore then hast thou brought me forth out of the womb?"

You can thank roving pieces of DNA that you were born rather than hatched from an egg. At least, that's what evolutionary biologist Vincent Lynch and his team at the University of Chicago tell us. *ScienceNews* magazine brings us the story about roving pieces of DNA that helped early mammals replace egg-laying in favor of giving birth to live young.

These "jumping genes" flipped the switch on thousands of genes, turning off ones that build hard eggshells and turning on genes that allow a fetus to develop in the uterus. Lynch said that the genes that turned on allowed the mother's body to recognize she is pregnant. It then suppressed her immune system so her body didn't sense the foreign DNA in the fetus and reject it. Animals that become pregnant, he continued, had the evolutionary advantage of carrying their developing young with them, rather than laying eggs in one spot where they'd be vulnerable to predators or unpredictable weather.

What the article fails to mention is that all of this – all of it – is sheer speculation. Can you imagine all of the *millions* of unaccounted-for changes that would need to take place to go from an egg-laying bird to a mammal that bears live young?

Just as you would expect from a team of evolutionary biologists, their research is built on the premise that evolution is a fact. They never even considered the more believable explanation that some creatures were designed to lay eggs and others were designed to give birth to live young.

Prayer: Heavenly Father, help Your children learn how to differentiate between truth and error, facts and speculation. Guide us into Your truth, I pray. Amen.

Ref: "Pregnancy in mammals evolved with help from roving DNA," *ScienceNews*, 1/29/15.

The Science of Parking Your Car

Isaiah 4:6
"And there shall be a tabernacle for a shadow in the daytime from the heat, and for a place of refuge, and for a covert from storm and from rain."

It's a sunny, sweltering summer day, and you're looking for a place to park your car. Since you know you'll be gone for several hours, you look for a nice shaded spot so your car will be more comfortable when you return. Or it's a blustery winter day with sub-zero temperatures. This time you park your car where there's as much sunlight as possible so that the sunlight streaming through your windows will heat your car's interior while you're gone.

What you are doing is practicing what scientists call behavioral thermoregulation. Through experience, we learn how to adjust our parking behavior to fit the climate conditions. But did you know that animals and plants also know how to adjust their bodies to help them stay warm or cool?

Insects and reptiles are particularly skilled at this. On a cold day, for example, an insect will adjust its position so that it intercepts as much sunlight as possible. Lizards and snakes bask in the sunlight on cold days, and on warm days seek relief in the shade or by burrowing underground.

Surprisingly, plants, too, know how to adjust their positions. The silk tree, for example, has leaves that it can move into three different positions depending on whether it needs to keep cool or gain heat.

Most likely, somebody taught you where to park your car for maximum comfort. Or you were intelligent enough to figure it out. But it could only have been God who taught the animals and the plants.

Prayer: Heavenly Father, since You have provided for the comfort and survival of the least of Your creatures, I know I can trust that You are providing for me as well. Amen.

Ref: R. Allen, *Bulletproof Feathers*, p. 121 (University of Chicago Press, 2010).

Dawkins Becomes a Creationist

Luke 24:31
"And their eyes were opened, and they knew him; and he vanished out of their sight."

When Creation Moments posted a story with the headline "Dawkins Becomes a Creationist," many evolutionists reacted in disbelief, saying that one of their own – namely, Richard Dawkins – could never become a creationist. And yet, a scientist and veterinarian by the name of Karen Dawkins was also an evolutionist ... until God changed her way of thinking.

In an interview with Creation Moments, she said, "As a product of the public school system, I always assumed that evolution was fact. Majoring in science in college, I began having my doubts about the scientific logic of evolution. The first organism I learned about that was not explainable by evolution was the woodpecker. There is no way it could have evolved with so many specialized organs. It had to have been created with all of its specialized organs all at one time. It still took me about fifteen years to come to the conclusion that God created the heavens and the Earth in six literal days."

She also told us, "Christianity explains the basics of science. Biology, chemistry, geology, physics and astronomy are all based on an orderly, predictable set of laws. And if life is ruled by these laws, then there has to be One who created those predictable sets of laws."

Today Karen Dawkins feels she has a more complete understanding of the sciences by the revealed Word of God. If such a radical change could happen to her, we can only pray that another evolutionist named Dawkins will come to see Christ as Lord and Creator before it's too late.

Prayer: Heavenly Father, I pray for evolutionists so that their turning to You will glorify Your Son and impact the lives of many others. Amen.

Ref: Karen Dawkins interview on file at Creation Moments. Karen E. Dawkins, who earned a Bachelor of Science degree as well as a Doctorate in Veterinary Medicine, is now teaching science as a homeschooling parent.

War Between Science and Faith?

1 Timothy 6:20
"O Timothy, keep that which is committed to thy trust, avoiding profane and vain babblings, and oppositions of science falsely so called."

Though evolutionists would have you believe there's a war between science and faith, the very idea of a conflict is preposterous. Creation Moments board member Dr. Donald Clark points out that many of the early modern scientists were Christians – like Galileo, Copernicus and Newton. Because of their faith, they believed in a God of order, and this drove them to look for the order in the universe in the first place.

But how things have changed since then! Today we are told that we must keep our faith out of classrooms and scientific institutions in order to do good science. Nothing could be further from the truth! As Dr. Clark writes, "This is a lie from the enemy of our soul to try to indoctrinate our children to not believe in God. How can you do good science if you do not believe that the stuff you are looking at is ordered and if you believe that it came about from chaos?"

Indeed, why perform any experiments at all if the next time you do it the results will be different? There is no point in it. So while many science professors say we must keep science and faith separate, they themselves can't believe this. Otherwise, they would never go into the laboratory.

According to Dr. Clark: "It all boils down to this: What do you put your faith in – science – which is man's understanding of the world – or God's Word? As for me and everyone at Creation Moments, our trust is in God's Word."

Prayer: Heavenly Father, I pray that You will protect our children and grandchildren from the lie of evolution and that You will guide them into Your truth. Amen.

Ref: Dr. Donald Clark, "War Between Science and Faith?" 4/30/12, Creation Moments online.

Six Hundred Years Before Darwin

2 John 1:7
"For many deceivers are entered into the world, who confess not that Jesus Christ is come in the flesh. This is a deceiver and an antichrist."

Most people consider Charles Darwin to be the father of evolution. The fact is, someone beat him to it by six hundred years. In the thirteenth century, a Muslim scholar by the name of Nasir al-Din Tusi proposed a theory of evolution that in many ways foreshadowed Darwin's.

Tusi's theory begins with a universe consisting of equal and similar elements. Those elements evolved into minerals and then to plants. From there they evolved to animals, and then to humans. He even put forward the idea that hereditary variability was an important factor for the biological evolution of living things.

This Persian scholar was born in the city of Tus in northeastern Iran in the year 1201. In addition to his work in biology, Tusi was an architect, astronomer, chemist, mathematician, poet, philosopher, physician, physicist and theologian. In all, he is responsible for 150 works, most of which are in Arabic.

Regarding evolution, he considered humans to be the "middle step of the evolutionary stairway." The final step, he wrote, can be achieved through man's will. Tusi was also the first to suggest that not only do organisms change over time but that the whole range of life has evolved from a point when there was no life at all. In other words, life from non-life.

There is one other point of similarity between Tusi's and Darwin's theories of evolution. Both theories are based on the faulty reasoning of man rather than the sure Word of God.

Prayer: Heavenly Father, I pray that You will help people come to a knowledge of the truth that's found in the pages of the Bible. Amen.

Ref: D. Green, *The Science Book*, p. 121 (DK Publishing, 2014).

Where the Battle Rages

Psalm 140:7
"O GOD the Lord, the strength of my salvation, thou hast covered my head in the day of battle."

Today's Creation Moment is taken from a viewpoint commentary written by Creation Moments Board Chairman Mark Cadwallader.

The date: October 31, 1517. The place: Germany. The event: Martin Luther nails his famous 95 Theses to the Wittenberg church door. Luther was primarily protesting the sale of indulgences, a direct attack on God's free gift of salvation.

As Luther put it, "If I profess with the loudest voice and clearest expression every portion of the truth of God except precisely that little point which the world and the devil are at that moment attacking, I am not confessing Christ, however boldly I may be professing Christ."

In addition to defending salvation by grace through faith, Luther also saw that the battle was raging over the Bible's account of creation. In his pithy way he told the skeptics, "When Moses writes that God created heaven and earth and whatever is in them in six days, then let this period continue to have been six days. But if you cannot understand how this could have been done in six days, then grant the Holy Spirit the honor of being more learned than you are."

Luther stressed the biblical view that the Word was God's instrument of creation, the Word which spoke things into existence and the Word which became flesh to win our salvation. In raising the text of Scripture to the level of highest authority in all things, Luther's was part of that view which gave birth to modern science – founded in the West by biblical creationists!

Prayer: Heavenly Father, with Satan attacking biblical creation, help me to be bold in defending the truth so others may know they can trust the Bible and turn from the lie of evolution. Amen.

Ref: Mark Cadwallader, "Where the Battle Rages," 9/27/11. Creation Moments online.

The Stupidity Virus

Ecclesiastes 10:3
"Yea also, when he that is a fool walketh by the way, his wisdom faileth him, and he saith to every one that he is a fool."

As we've mentioned on previous Creation Moments programs, our bodies contain trillions of bacteria, viruses and fungi, most of which are harmless. But researchers have now discovered a virus that infects human brains and makes us more stupid. No, we're not making this up.

The ATCV-1 virus is ordinarily found in green algae that inhabit rivers and lakes. So far, scientists have no idea how the micoorganism gets inside the throats of some humans. But this much they do know – it is not contagious and does not adversely affect a person's health. It just makes them 10 percent more stupid.

The irony is that the scientists at Johns Hopkins Medical School and the University of Nebraska stumbled upon this discovery by sheer dumb luck. They were actually working on an unrelated study into throat microbes. Virologist Dr. Robert Yolken tells us: "This is a striking example showing that the 'innocuous' microorganisms we carry can affect behavior and cognition."

Of the 90 participants in the study, 40 tested positive for the algae virus. Those who tested positive performed worse on tests designed to measure the speed and accuracy of visual processing. They also achieved lower scores in tasks designed to measure attention.

Like a virus that affects our behavior, the sin nature we're all born with makes us "stupid" in the knowledge of God. But God has overcome our fallen sin nature and makes us righteous through the blood of Jesus, poured out for us on the cross.

Prayer: Heavenly Father, thank You for the Holy Spirit who helps me avoid stupid actions and stupid thinking. Help me follow His leading so I may behave in a way worthy of my calling. Amen.

Ref: "Virus that 'makes humans more stupid' discovered," *The Independent*, 11/9/14. "American Researchers Discover 'Stupidity Virus'," *Newsweek*, 11/10/14.

Knocking Evolution Over with a Feather

Deuteronomy 32:11
"As an eagle stirreth up her nest, fluttereth over her young, spreadeth abroad her wings, taketh them, beareth them on her wings:"

The foundation of evolution is so weak, the flight feather of a bird can send it crumbling to the ground. Perhaps no one is better equipped to speak on the amazing design of flight feathers than Dr. Stuart Burgess, the engineer who designed the solar array deployment system on the European Space Agency's earth observation satellite. In addition to his engineering work at Bristol University, he is the author of *Hallmarks of Design: Evidence of Purposeful Design and Beauty in Nature.*

In this book he points out that the flight feather is a masterpiece of design and is one of the most efficient structures known to man. The hooks and barbules of a feather are arranged so that they prevent air from passing through them when the wing is pushing downwards, but they allow air to pass through them when the wing is being pulled upwards. This feature, he writes, enables the bird to maximize the efficiency of flapping by only allowing the wing to push down on the air.

While evolutionists admit that the feather is an amazing structure, they insist that it evolved by a long series of genetic mistakes. But Dr. Burgess makes it clear that this is just impossible. The interlocking hooks and barbules must all be in place simultaneously for the feather to function.

To borrow a phrase from the Intelligent Design folks, the feather is irreducibly complex. All it takes is a single feather to show that the theory of evolution is a theory that just won't fly.

Prayer: Lord Jesus, whenever I see a bird in flight, remind me that it is You who gave them all the right equipment so they can take wing and fly. Amen.

Ref: S. Burgess, *Hallmarks of Design*, pp. 38-39 (Day One Publications, 2002).

Amazing Underwater Eyes

Psalm 94:9
"He that planted the ear, shall he not hear? he that formed the eye, shall he not see?"

We love it when one of our listeners brings something of interest to our attention that we can pass along to others. A listener sent us a link to an article at the *National Geographic* website on little-known facts about the eyes of animals. Since *National Geographic* treats evolution as fact, no mention was made that God created the eyes of these animals. But we are sure that our listeners will be quick to give credit where credit is due!

When asked which creature has the most eyes, the writer replied that the world record holder is a type of mollusk called chiton. This ocean dweller has thousands of eyes embedded in shells on their backs. Most scallop species also have up to hundreds of eyes, as do giant clams. What's more, certain types of tubeworms called sabellids and serpulids have hundreds of small, compound eyes on their feeling appendages.

When asked which animal has the most advanced vision system, the article gave the nod to another underwater creature – the mantis shrimp. Their stalked eyes have three separate regions, and they are even able to see ultraviolet and circularly polarized light. Compared to our own three photoreceptors, mantis shrimp have up to sixteen! These crustaceans may even use their photoreceptors to recognize colors right in the eye, rather than processing them in the brain, as we do.

If you come across anything in a book or at a website that you feel would be of interest to other listeners, let us know by sending an e-mail to info@creationmoments.com.

Prayer: *Heavenly Father, thank You for giving me eyes to see and especially for giving me the eyes to see spiritual things that lead to eternal life. In Jesus' Name. Amen.*

Ref: L. Langley, "Ask Your Weird Animal Questions: Animal Eyes," *National Geographic*, 7/14/14.

The Greatest Mass Murderer of All Time

John 8:44a
"Ye are of your father the devil, and the lusts of your father ye will do. He was a murderer from the beginning..."

Who do you think was the greatest mass murderer the world has ever seen? Adolf Hitler? Not even close. According to Creation Moments Board Chairman Mark Cadwallader, the greatest mass murderer was China's Mao Tse Tung. During his political reign, Mao was responsible for murdering 77 million people while enforcing his "Great Leap Forward," and he "purged" millions more during what he called "The Cultural Revolution."

Interestingly, Mao is reported to have said the following: "Chinese socialism is founded upon Darwin and the theory of evolution." Now, who's the runner-up to Mao? No, it's still not Hitler. The second greatest mass murderer was Joseph Stalin, who similarly saw the value of evolution to his plans for a collectivist state. He used purges, starvation and prison camps to force people out of their own property and into acceptance of the ideology where God is replaced by government. He was responsible for the murder of up to 40 million people.

The presumption that evolution is a scientific fact, disproving God and validating the philosophy of "might makes right," has justified terrible atrocities in the post-Darwin twentieth century.

Darwinism isn't just responsible for the physical death of millions, it has led to the spiritual death of many millions of people who prefer to live in a universe where God does not exist. Thankfully, God is opening the eyes of many to the truth found in the pages of the Bible – that we are not the product of evolution but were created by God.

Prayer: Father, I have friends who are deceived by the lie of evolution. I pray You will use me to help some of them find the narrow path that leads to eternal life. Amen.

Ref: Mark Cadwallader, "Who Is the Greatest Mass Murderer of All Time?" 7/24/13. Creation Moments online.

There's Something Fishy About This Car

Isaiah 53:2b
"...he hath no form nor comeliness; and when we shall see him, there is no beauty that we should desire him."

Why in the world would engineers at Mercedes-Benz and DaimlerChrysler turn to a fish built like a box for inspiration on automotive design? After all, automakers want their vehicles to travel just as efficiently through air as sharks swim through water. But the boxfish is anything but streamlined.

The fact is – the cube-shaped boxfish slips through water with unparalleled efficiency. As the journal *Science* put it, "One look at the aptly named boxfish, and you might expect it to swim as well as a barn would fly." Creationist Donald DeYoung points out in his book *Discovery of Design*, "Observations show that the boxfish swims easily and safely, even in turbulent water. Self-correcting vortices of current develop around its body and effectively cancel out the buffeting forces of surging water."

DeYoung pointed out that though the automotive engineers had gone to the aquarium to design a car patterned after the sleek shark, the vehicles DaimlerChrysler ended up basing their automobile designs on was the boxfish. And it wasn't just the fish's shape that they copied. By patterning the door panels after the hexagonal skin pattern of the fish, the cars were found to excel in safety, comfort, maneuverability and environmental friendliness. The compact cars even boasted excellent gas mileage to as high as 70 miles per gallon!

Only our glorious Creator could have designed a clunky-looking fish that can help automobile designers develop safer and more fuel-efficient cars!

Prayer: Lord Jesus, thank You for creating a world filled with magnificent creatures that not only help mankind but also bring glory to You! Amen.

Ref: Donald DeYoung, *Discovery of Design*, pp. 74-75 (Master Books, 2009).

Tall, Fat and Upside-Down

Luke 6:44
"For every tree is known by his own fruit. For of thorns men do not gather figs, nor of a bramble bush gather they grapes."

If you want to snack on fruit from a ten-story-high Grandidier's baobab tree, you'd better bring a very tall ladder. That's because there are no branches to climb until you get to near the very top of the tree. And you can forget about wrapping your arms and legs around the trunk. The baobab's massive cylindrical trunks measure up to 10 feet across, and are quite smooth.

The Grandidier's baobob is probably the strangest and most magnificent of the eight species of baobab, and it is found nowhere else on earth than Madagascar. Lack of water can sometimes be a problem for plants in Madagascar, so the baobab overcomes this by storing water within the fibrous wood of the trunk. It is so effective at storing water, the tree's diameter actually fluctuates with rainfall.

Other species of baobab tree – like the ones in Africa – look as if they were built upside down – with their roots at the top of the tree. African baobabs are also the fattest trees in the world. Some have a "waist size" of more than 165 feet, which is bigger around than their height. Though the tree trunks look solid, elephants know they can gouge into the wood with their tusks to enjoy a nice drink of water. Baobabs can also grow to a great age – some are said to have been alive at the time of Christ.

Baobab trees look very different than any other trees on our planet and demonstrate that there is no end to the creativity of God.

Prayer: Heavenly Father, there is no limit to Your creative power! Even the trees lift up their branches in praise to You! Amen.

Ref: "Overweight and Upside Down," *1000 Wonders of Nature,* p. 243 (Readers Digest Assn., 2001). "Grandidier's baobab," Wildscreen Arkive.

The South-Pointing Carriage

1 Samuel 17:22
"And David left his carriage in the hand of the keeper of the carriage, and ran into the army, and came and saluted his brethren."

Numerous Chinese legends refer to a mysterious carriage on which a mechanical rider rode. Its arm was outstretched and continuously pointed south no matter which way the carriage turned. How did it accomplish such a feat? If you're guessing that magnets had something to do with it, you'd be wrong. Actually, the south-pointing chariot was invented about 800 years before the first navigational use of a magnetic compass.

In the 1960s, Dr. Joseph Needham discovered that the carriage dated back to the third century AD. He also learned that the device did not use magnets. The carriage was actually a highly sophisticated automaton that has since been described as the world's "first cybernetic machine."

So how did the figure know how to keep pointing south? It used a complex grouping of gears similar to the differential gears found in modern cars. This system of interconnected gears allows the wheels of a car turning a corner to move at different speeds, with the "outer" wheel traveling farther than the other. The arm of the figure on the carriage was aimed southward by hand at the start of a journey. After that, whenever the carriage turned, the mechanism automatically rotated the arm to keep it pointed to the south.

Over the years, we have told you about many such ancient devices, revealing that ancient peoples were much more intelligent than modern man gives them credit for. This should come as no surprise to those of us who believe the Bible.

Prayer: Heavenly Father, though evolutionists want us to believe that humans developed slowly over millions of years, the Bible tells us that mankind was intelligent and skilled right from the beginning. Amen.

Ref: P. James and N. Thorpe, *Ancient Inventions*, pp 140-142 (Ballantine Books, 1994).

Don't Have a Cow, Man!

Jeremiah 13:23
"Can the Ethiopian change his skin, or the leopard his spots? then may ye also do good, that are accustomed to do evil."

Many thousands of people have visited the Creation Moments website after seeing a photo we posted at the popular WND.com website. The photo shows a cow with spots on its hide that look exactly like a map of the world. We asked readers if they thought the map on the cow was just a lucky arrangement of spots or if the photo was the product of a Photoshop artist.

Obviously, everyone knew that the cow's spots were put there by a graphics designer. But then we asked evolutionists to answer this question: If you agree that the cow's spots were designed, why do you say that the actual cow was not designed? After all, a living, breathing cow is infinitely more complex than the arrangement of spots on its hide.

Evolutionists and atheists, of course, acknowledge that the pattern of the cow's spots was designed, but they will *never* admit that the cow itself was designed. That's because they are held captive by their faith built on the premise that there is no Designer. Though they can see the complexities of nature all around them, they say that everything was the result of mindless, natural processes.

And that's why Darwinists have a cow whenever they hear a Bible-believing Christian say that things which appear to be designed actually *were* designed! If they weren't so biased against God, they'd know that the cow's spots in the photo reveal the incredibly huge blind spot in their own minds.

Prayer: Heavenly Father, I pray that You will open the eyes of evolutionists to the truth that Your creation is so complex that it requires the existence of a Designer. Amen.

Ref: http://www.creationmoments.com/node/5205.

In the Blink of an Eye

Proverbs 20:12
"The hearing ear, and the seeing eye, the LORD hath made even both of them."

Did you know that the intermittent windshield wiper on your car was discovered in the blink of an eye? According to the book *Inspired Evidence: Only One Reality*, here's the story of how it came about.

Drivers today take the intermittent windshield wiper for granted. After all, every car has them. But before 1969, drivers weren't so lucky. Their wipers had just two speeds – fast and slow. And most of the time, neither speed was exactly what the driver wanted.

Back in the early 1960s, Robert Kearns, an engineering professor at Wayne State University in Detroit, Michigan, was having trouble seeing as he was driving his car on a misty night. He had nearly been blinded on his wedding night a decade earlier when the cork from a bottle of champagne shot into his left eye. With his limited vision and the mist on his windshield, he started thinking about the human eye which has its own kind of wiper, the eyelid, that automatically closes and opens intermittently as needed every few seconds.

In 1963, Kearns built his first intermittent wiper system using off-the-shelf electronic components. In his design, the interval between wipes was determined by the rate of current flow into a capacitor. When the charge in the capacitor reached a certain voltage, the capacitor was discharged, activating the wiper motor for one cycle. Ford incorporated intermittent wipers on the company's Mercury line, beginning with the 1969 models.

The intermittent windshield wiper – like many other practical inventions – was inspired by the designs God put into His creation.

Prayer: Lord Jesus, Your creation is filled with so many examples of ingenious design, I can't even begin to count them all! I praise Your most holy name! Amen.

Ref: J.V. Vette and B. Malone, *Inspired Evidence: Only One Reality*, June 24 (Search for the Truth Publications, 2011).

You Don't Use Science to Test Scripture!

Matthew 24:35
"Heaven and earth shall pass away, but my words shall not pass away."

How often have you heard people say that the Bible must be wrong because science disagrees with it? That attitude certainly isn't new. It was common in 1932 when Dr. Harry Rimmer wrote his classic book, *The Harmony of Science and Scripture*. Even back then, Bible skeptics were using science to test Scripture. But as Dr. Rimmer pointed out, when science has matured to the point of infallibility, then and only then could it be used to test Scripture.

To emphasize that point, he mentioned a list of fifty-one scientific facts published by the French Academy of Sciences in 1861, all of which contradicted some statement from Scripture. When Rimmer sat down to write his book some seventy years later, he noted that the Bible hadn't changed in all that time, but "the knowledge of science has so vastly increased that there is not a living man of science today who holds one of those fifty-one so-called facts that were at one time advanced in refutation of the inspiration of the Scripture."

The modern craze to test Scripture by science reverses the natural order, he said. "After twenty-five years of research on this subject, we are willing to admit that where science and the Bible are in utter harmony, that agreement establishes the certainty *of that science*."

To Dr. Rimmer, using science to test Scripture was like a seven-year-old boy telling a seventy-year-old man how to grow up. How right he was! You just don't use science to test Scripture!

Prayer: Heavenly Father, unlike science textbooks that are constantly being revised, Your Word never needs to be updated. Amen.

Ref: Harry Rimmer, *The Harmony of Science and Scripture*, pp. 58-59 (Eerdmans Publishing, 22nd printing, 1973).

Why Did God Give Us Fingerprints?

Genesis 27:21
"And Isaac said unto Jacob, Come near, I pray thee, that I may feel thee, my son, whether thou be my very son Esau or not."

Criminals who have been linked to a crime through their fingerprints may not be happy that they were born with a unique pattern of "grooves" on their fingertips. But we can all be glad that God gave us fingerprints because they greatly improve our sense of touch.

Scientists in France performed a series of experiments with artificial fingertips made of rubber-like sensors. They then compared the sensitivity between these grooved artificial fingertips and a smooth skin-like material to see if the grooved fingertips made a difference in the sense of touch. What they discovered surprised them. The grooved fingertips produced vibrations up to 100 times stronger than the smooth material when sliding against a slightly rough surface.

The researchers concluded that these increased vibrations provide us with a greater ability to detect textures. When rubbing your fingers across a textured surface, your fingerprints amplify vibrations that stimulate the nerve endings in your skin. This then allows us to identify objects by touch.

Knowing that we need our sense of touch to work from every direction, our Creator designed our fingerprints to appear in elliptical swirls. This loop design ensures that some ridges are always brushing perpendicular to a surface, no matter the orientation of our fingertips.

Such research may help scientists design prosthetic hands with enhanced tactile feedback. Once again, when science copies designs found in living things, it is copying the designs put in living things by their Creator, whether they recognize it or not!

Prayer: Lord Jesus, thank You for all of my senses, including the sense of touch. Amen.

Ref: L. Zyga, "Why Do We Have Fingerprints?" PhysOrg, 4/4/09. "The Role of Fingerprints in the Coding of Tactile Information Probed with a Biomimetic Sensor," J. Scheibert, S. Leurent, A. Prevost, and G. Debregeas (13 March 2009) Science 323 (5920), 1503. DOI: 10.1126/science.1166467.

The Silence of the Owls

Isaiah 34:15a
"There shall the great owl make her nest, and lay, and hatch, and gather under her shadow:"

What makes owls so good at catching prey as they fly through the night sky? Part of the credit obviously goes to their amazing eyes that are able to see with such clarity in low-light conditions. But owls also have another design feature that allows them to sneak up on their prey without being noticed. Owls, you see, were designed to fly in virtual silence.

The authors of the book *Discovery of Design* point out that owls have an uneven forward fringe on their wings. Unlike the sharp, well-defined edge on the wings of most birds, the uneven fringe decreases air turbulence and produces less noise. In addition, the feathers covering the owl's wings, body and legs are velvety soft. This helps to dampen and absorb the sound of rushing air.

Airplane designers are now exploring these features to create quieter military and commercial aircraft. Thanks to the owl, engineers are looking into a retractable brush-like fringe for airplane wings and a velvety coating on the landing gear.

In the book's introduction, the authors point out that inventors and design engineers frequently look to nature for inspiration. But as creationists, they emphasize that the designs found in nature are *not* the product of evolution. Rather, the designs were embedded in the material universe by supernatural acts of creation. The purpose of these designs was not only for the benefit of living things but also so they could be discovered and put to use for the welfare of mankind.

Prayer: Heavenly Father, the creation not only inspires designs that benefit mankind, they inspire us to worship our Creator! I am filled with awe as I learn more about Your creation. Amen.

Ref: D. DeYoung and D. Hobbs, *Discovery of Design: Searching Out the Creator's Secrets*, pp. 9-10, 66-67 (Master Books, 2012).

Fossil Jaw Not Evidence for Evolution

2 Thessalonians 2:11
"And for this cause God shall send them strong delusion, that they should believe a lie:"

Evolutionists are rejoicing over a fossil jaw found in the Ledi-Geraru research area of Ethiopia. They say it pushes back evidence for early humans to 2.8 million years ago and predates all previously known fossils of mankind's lineage by about 400,000 years. The fossil which has caused such excitement consists only of the left side of a lower jaw and five teeth.

Research team member William Kimbel would have us believe that "the Ledi jaw helps narrow the evolutionary gap between *Australopithecus* and early *Homo*. It's an excellent case of a transitional fossil in a critical time period in human evolution."

But biochemist and creation scientist Dr. Donald Clark strongly disagrees. "You would think that they've got definitive proof that this is in the human line. It just amazes me that they can jump to conclusions based on very minimal evidence." Dr. Clark also said, "Everything that [the research team] would say about an ape or a chimp lower jaw bone ... this particular specimen has. The jaw is U-shaped like an ape's would be. The teeth are very flat; they are not sharp and pointed like our teeth would be. The reason why they say [the fossil is in the human lineage] is because the flattened teeth are a little bit smaller than what you would find on a normal ape today. That, to me, is not definitive."

How true! If anything, this fossilized jaw only shows how weak the case for ape-to-human evolution really is.

Prayer: Heavenly Father, I pray that You will awaken many of those who are now caught in the delusion of evolution. In Jesus' Name. Amen.

Ref: "Discovery of 2.8-million-year-old jaw sheds light on early humans," *ScienceDaily*, 3/4/15. Dr. Donald Clark interview on Broken Road Radio, 3/10/15. http://brokenroadradio.com/?p=3508

Cosmologist Paul Davies on Faith

Colossians 4:6
"Let your speech be alway with grace, seasoned with salt, that ye may know how ye ought to answer every man."

I could easily fill up today's entire Creation Moments broadcast by listing the scientific accomplishments and affiliations of Paul Davies. His research interests span the fields of cosmology, quantum field theory and astrobiology.

An opinion piece he wrote in 2010 for the *New York Times* angered many evolutionists when he argued that the faith scientists have in the immutability of physical laws has origins in Christian theology. Davies responded to their criticisms by saying, "I was dismayed at how many of my detractors completely misunderstood what I had written. Indeed, their responses bore the hallmarks of a superficial knee-jerk reaction to the sight of the words 'science' and 'faith' juxtaposed."

Well, what Davies wrote in London's *Guardian* newspaper in 1982 would have angered them even more. "Something strange is happening to the universe," he wrote. "In the words of the astronomer Fred Hoyle, it is as though somebody has been monkeying with the laws of nature. A once-popular argument with theologians was to point out how astonishingly well ordered the universe is, how harmoniously its components dovetail together. All this, it was reasoned, must be the result of design, and therefore evidence of a Great Designer. Yesterday's theologians would have been delighted with today's discovery of just how delicately balanced the cosmic order turns out to be."

Though Davies freely admits he is "not a religious man," he brings a calm and mature voice to the debate over origins that other evolutionists would be wise to follow.

Prayer: Heavenly Father, I pray that the debate over origins would take place in a dialog that is free from emotional outbursts and name-calling. Let our discussions generate more light and less heat. Amen.

Ref: A. "The Great Conundrum in the Sky," *The Guardian*, 9/2/82, p. 17. "Taking Science on Faith," *New York Times*. 10/2/10. Wikipedia entry on "Paul Davies."

The Anti-Theory of Evolution

Colossians 2:8
"Beware lest any man spoil you through philosophy and vain deceit, after the tradition of men, after the rudiments of the world, and not after Christ."

When a well-respected scientist says anything that is critical of Darwinism, two things are bound to happen. First, evolutionists will put pressure on the scientist to retract his comments. Second, creationists will quote the scientist to show that even some scientists understand the weaknesses of evolution. This is what we are doing on today's broadcast.

In his book *A Different Universe,* Nobel Laureate Robert Laughlin, a professor of physics and applied physics at Stanford University, wrote that evolution is anti-science and actually impedes scientific investigation. Laughlin wrote he is concerned that much "present-day biological knowledge is ideological" and cannot be tested.

Here is an excerpt of what he wrote: "Evolution by natural selection, for instance, which Charles Darwin originally conceived as a great theory, has lately come to function more as an anti-theory, called upon to cover up embarrassing experimental shortcomings and legitimize findings that are at best questionable.... Your protein defies the laws of mass action? Evolution did it! Your complicated mess of chemical reactions turns into a chicken? Evolution! The human brain works on logical principles no computer can emulate? Evolution is the cause!"

Now, let me point out that Creation Moments is not saying that Dr. Laughlin rejects evolution. He is merely saying what we have been saying for many years. Evolution is an ideology that is simply assumed to be true. Rather than being a scientific theory, it is an anti-theory that clearly lies outside the realm of true science.

Prayer: Heavenly Father, evolutionists claim that creationists are anti-science. Help them to see that they are the ones who have abandoned true science in favor of a godless ideology. Amen.

Ref: Robert B. Laughlin, *A Different Universe,* pp. 168-169 (Basic Books, 2005). J. Bergman, *The Dark Side of Darwin,* p. 53 (Master Books, 2011).

Smallest Life Form Discovered

Mark 8:7-8
"And they had a few small fishes: and he blessed, and commanded to set them also before them. So they did eat, and were filled: and they took up of the broken meat that was left seven baskets."

A bacterium has recently been photographed that is so tiny, researchers have debated about whether it could even exist. The bacterium has a volume of only .009 cubic microns. To put that into perspective, a micron is one millionth of a meter. This means that more than 150,000 of them could fit on the tip of a single human hair.

Despite their small size, these bacterium have everything they need to survive and reproduce. And what they don't have, they borrow. For instance, these bacteria have tiny string-like structures called pili. Scientists suspect the bacteria use these pili to link up with other microbes so they can borrow key nutrients from them.

Now, don't let evolutionists tell you it would be easy for evolution to produce a life form as tiny as this bacterium. If anything, its small size speaks loudly of an incredibly intelligent Creator. Consider, for example, the phone in your purse or pocket. Today's typical smartphone is so technologically advanced, it has far more computing power than computers from the past that filled up entire rooms. And they can do far more than make phone calls. Hey, we call them *smart* phones for a reason!

Little things can accomplish astonishing things in the hands of a great God. Jesus took a few small fish and used them to feed thousands. God took a little boy and used him to vanquish a giant. So don't ever think you are too small or insignificant to do great things with God.

Prayer: Heavenly Father, from the tiniest life form to the greatest galaxy, You created them all! Amen.

Ref: E. Brodwin, "Scientists have finally found and photographed the tiniest life on Earth," Business Insider/Yahoo News, 3/3/15.

The Bird Dung Spider

Romans 5:8
"But God commendeth his love toward us, in that, while we were yet sinners, Christ died for us."

Though it must surely be one of the most disgusting creatures in God's creation, the bird dung spider is yet another beautiful example of God's design.

Not only does the bird dung spider have a body covered with unattractive warts, the spider often produces a white thread and sits on it, looking almost exactly like bird poop that has fallen on a leaf.

Despite its unattractive appearance, the spider's unique camouflage is a masterpiece of well-thought-out design. First off, by looking like bird poop, the spider is of no interest to a wide variety of potential predators. At the same time, its camouflage brings a steady flow of dung-loving insects within reach of the slow-moving spider. On top of that, the spider is able to attract male moths by mimicking the sex pheromone released by female moths. As one website put it, "Once the male moth gets a whiff of the scent, it will fly towards the smell expecting to find a female moth. Imagine the moth's surprise when the motionless bit of bird dropping comes alive! The spider will use its front legs to grab the moth and then it will devour it on the spot."

Just as the bird dung spider is unattractive to us, so, too, are we unattractive to God in our fallen state. Nevertheless, while we were yet sinners and deserving of God's wrath, Christ died for us. God saved us not because of any beauty He saw in us but because He loved us. By shedding His blood for us on the cross, He transforms us into a new and beautiful creation!

Prayer: Lord Jesus, by shedding Your blood for me on the cross, You have transformed me into a new and beautiful creation! Amen.

Ref: "Bird-Dung Spider," SpiderVista online. "Bird-dropping Spider - the Mimic Master," http://www.brisbaneinsects.com/brisbane_orbweavers/BirdDroppingSpider.htm.

Another Attempt to Erase Jesus from History

Psalm 50:22
"Now consider this, ye that forget God, lest I tear you in pieces, and there be none to deliver."

When most of us were in school, everybody knew that the terms "BC" and "AD" meant Before Christ and Anno Domine, meaning "in the year of our Lord." But those terms are disappearing as authors now prefer the terms CE and BCE, which are abbreviations for "Common Era" and "Before the Common Era."

As the online encyclopedia Wikipedia points out, "since the later twentieth century, use of CE and BCE has been popularized in academic and scientific publications." It is also being used "by publishers wishing to emphasize secularism and/or sensitivity to non-Christians." In other words, they are changing the terms to erase Jesus from history under the guise of not wishing to offend anyone. The secularists, on the other hand, have no qualms about offending Christians.

This is what secularism does. It obscures the truth and even revises history in subtle or overt ways. Applied chemist and creation scientist Mark Cadwallader says that the use of CE and BCE is "clearly another example of the assault on Christianity. Radical atheists," he says, "have been agitating for freedom from any reference to Christianity and especially its founder. They will disguise and revise history itself to obscure the truth of Jesus Christ."

Here at Creation Moments, we believe that truth will ultimately prevail, but we must all do our part. That's why we encourage you to do everything you can to make all people, young and old, aware of the truth and consistency of a biblical worldview. And keep on using BC and AD!

Prayer: Heavenly Father, though sinful man is trying to erase Your Son from history, they will not succeed. Someday every knee will bow and every tongue will confess that Jesus is Lord! Amen.

Ref: Mark Cadwallader, "Another Attempt to Erase Jesus from History," 2/27/12. Creation Moments online.

Satellite Image of the Exodus?

Acts 7:36
"He brought them out, after that he had shewed wonders and signs in the land of Egypt, and in the Red sea, and in the wilderness forty years."

The 40-year-long journey of Moses and the Israelites from Egypt to the land of promise continues to fascinate movie makers. From Cecil B. DeMille's 1923 silent film "The Ten Commandments" to the more recent "Exodus: Gods and Kings," the story of the Exodus never fails to excite our imagination. Even so, there are Bible scholars out there who are saying the Exodus never really happened. Some even insist that the Israelites never set foot in ancient Egypt.

Archaeologists have shown, however, that the biblical account is historically accurate. And satellite image analyst George Stephens even claims that the route of the Exodus can still be seen today through the use of infrared technology.

According to the book *The Stones Cry Out,* Stephens studied SPOT satellite imagery of Egypt, the Gulf of Suez, the Gulf of Aqaba and portions of Saudi Arabia. On the images, taken at an altitude of 530 miles, he claims he saw evidence of ancient tracks made by "a massive number of people" leading from the Nile Delta that eventually wound up in the Sinai peninsula. He said he even saw traces of "very large campsites" along the trail.

While it is impossible, of course, to know if these tracks were made by the Israelites during the exodus of Moses' time, the satellite images do demonstrate that large numbers of people could make long journeys through that inhospitable region and on the same route taken by the Israelites.

Prayer: Heavenly Father, thank You for raising up archaeologists who are confirming the historical accuracy of the Bible. Amen.

Ref: Larry Williams, *The Mountain of Moses* (Wynwood Press, 1990) "Epilogue" as cited in *The Stones Cry Out*, Randall Price, pp. 136-137 (Harvest House Publishers, 1997).

Evolutionists Still Defend Abiogenesis

John 1:3-4
"All things were made by him; and without him was not any thing made that was made. In him was life; and the life was the light of men."

Though Louis Pasteur proved that life comes only from life, the myth that life can come from non-life – or abiogenesis – is still defended by evolutionists today. They now claim that while Pasteur disproved spontaneous generation, he did not prove that life comes only from life.

Confused? Here is how one evolutionist explains it: "Spontaneous generation held that life in its present form today could form from non-life, and did so all the time – for instance, aphids sprang from dew on plants, maggots emerged from rotting meat, and mice were created from wet hay. In 1859, Louis Pasteur ... proved definitively that life does not spring, fully formed and unbidden, from any recipe of inorganic or dead organic matter."

Notice the words "fully formed" in that statement? Like most evolutionists today, he believes that life came from non-life, but such life came into being not at a fully formed *creature* level but at a *chemical* level. According to the evolutionist, "Abiogenesis predicts that ... the constituent components of life can self-arrange given certain conditions" and that "there is some point in Earth's early history wherein a chemical chain reaction went runaway and breached the fuzzy barrier between chemistry and biology."

Though evolutionists admit they have zero evidence for this, they still cling to such a rescuing device because evolution and their atheistic worldview depend on it. Unlike evolutionists, though, creationists can hang their hat on the established Law of Biogenesis – that life comes only from life!

Prayer: Heavenly Father, I thank You for creating life and especially for the gift of eternal life, earned for me by Your Son, Jesus Christ. Amen.

Ref: Jason Thibeault, "Abiogenesis is not spontaneous generation. Period." Posted 6/25/10 on freethoughtblogs.com.

Doctor Moses

Leviticus 13:45
"And the leper in whom the plague is, his clothes shall be rent, and his head bare, and he shall put a covering upon his upper lip, and shall cry, Unclean, unclean."

Did Moses know that germs – invisible to the naked eye – were the cause of disease thousands of years before the invention of the microscope? American evangelist, creationist and author Harry Rimmer thought it was possible. In his 1936 book, *The Harmony of Science and Scripture,* Rimmer wrote that the custom of wearing a protective mask in the hospital is a modern method of preventative medicine based upon the certainty that pathogenic organisms cause disease in the human body.

"How, then," he asked, "can we account, on purely human grounds, for the use of this same method for the prevention of infection in the days of Moses?" Rimmer then refers the reader to Leviticus 13:45 where "we have a record of a contagious disease.... The law of God, as given through Moses, contained the injunction that the infected man must bind a cloth across his upper lip, *exactly as the physician in the hospital wears the mask today!"*

Of course, Moses couldn't have known anything about the microorganisms that spread disease from one person to another, but God certainly knew all about germs! Rimmer concluded: "If Moses really spoke by inspiration of the Spirit of God and transmitted to us only those things which he in turn had received, we have a sensible explanation for this marvelous anticipation of modern wisdom in this ancient book."

While Creation Moments does not agree with Dr. Rimmer's position on the gap theory, we wholeheartedly agree that the Bible contains many scientifically sound truths we are only learning about today.

Prayer: Heavenly Father, You saved the Israelites who looked at the brass serpent Moses raised up on a pole. So, too, did You save me when I looked in faith at Your Son who was raised up on the tree of Calvary. Amen.

Ref: Harry Rimmer, *The Harmony of Science and Scripture*, pp. 101-102 (Eerdmans Publishing, 22nd printing, 1973).

Poetic Verses on Multiverses

Ephesians 4:14a
"That we henceforth be no more children, tossed to and fro, and carried about with every wind of doctrine..."

The idea of multiverses – that our universe is one of many – has long been used in science fiction stories. But now, atheists and evolutionists are depending on it to be true! They're saying this to dispose of the obvious fact that we live in a finely tuned universe that bears the stamp of our Creator's design. We are lucky to inhabit just the right universe, they say. This must surely be the pinnacle of wishful thinking! In his poem "Multiverses to the Rescue!", poet and creationist Tom Graffagnino points out that evolutionists desperately hope that multiverses will come to the rescue of evolution!

When we analyzed the figures,
We thought Chance and Time would do.
But we couldn't fake the numbers,
So we dreamed up something new!

Hmm, we pondered, let's pretend here,
Make believe – that's better still!
Multiverses! That's the answer!
Oh my, yes, that fits the bill!

We thought Chance 'n Time could do it,
But the notion crashed and burned.
Multiverses to the rescue!
My-oh-my! So much to learn!

Though we've never really seen one,
And we don't know if they're there;
Still by Faith we just believe it,
Gospel Science fills the air.

Endless Mythic Multiverses,
Brooms 'n Buckets, come alive!
Hocus-pocus is our focus,
Fantasy in overdrive.

Yes, the Scientism Genie
Has at last moved in to stay!
Glory, glory, Hallelujah!
Science fiction saves the day!

Bowing to Almighty Matter,
Yes, our hope's in Science now!
Naturalism is our Dogma,
Mind of Man our Sacred Cow.

Prayer: Heavenly Father, as I watch evolutionists fabricating universes out of their own imagination, I thank You for planting my feet in reality. Amen.

Ref: "Multiverses to the Rescue!" by creationist and artist Tom Graffagnino. Used with permission. For more information about Graffagnino and his work, visit his website: http://www.withoutexcusecreations.net/.

The Unsolved Origin of Species

Genesis 1:25
"And God made the beast of the earth after his kind, and cattle after their kind, and every thing that creepeth upon the earth after his kind: and God saw that it was good."

While most evolutionists today would deny what I am about to say, many prominent evolutionists would agree that Darwin's *Origin of Species* utterly failed to explain the origin of species.

Let me begin by making it clear that all of the people I'm mentioning today are evolutionists who were quoted in the book *The Great Evolution Mystery* by evolutionist Gordon Taylor. In the chapter dealing with the name of Darwin's famous book, Taylor quotes Professor Ernst Mayr of Harvard, who said, "The book called *The Origin of Species* is really not on that subject." His colleague, George Gaylord Simpson, said: "Darwin failed to solve the problem indicated by the title of his work."

Also quoted is British geneticist William Bateson, who wrote that the "essential bit of the theory of evolution which is concerned with the origin and nature of species remains utterly mysterious."

Taylor notes that the issue Darwin failed to deal with was speciation – that is, the evolutionary process by which new biological species arise. As creationists have rightly pointed out, Darwin dealt only with the *survival* of the species, not the *arrival* of the species. As Professor Hampton Carson of Washington University in St. Louis pointed out, speciation is "a major unsolved problem of evolutionary biology."

And it remains unsolved to this day. While neo-Darwinists continue to search for the answer, those of us who believe what the Bible teaches already know where the different kinds of animals came from. It's all there in the first chapter of Genesis!

Prayer: Oh Lord, if it were not for Your grace, I could have ended up believing the lie of evolution. Thank You for helping me see that evolution is an emperor without any clothes. Amen.

Ref: Gordon Rattray Taylor, *The Great Evolution Mystery*, pp. 140-141 (Harper & Row, 1983). Taylor (1911-1981) was a popular British journalist and evolutionist who was critical of some elements of neo-Darwinism.

Designed to Stand

Ephesians 6:14
"Stand therefore, having your loins girt about with truth, and having on the breastplate of righteousness;"

Try as they might, no evolutionist can successfully explain how a four-legged ape-like creature evolved into a two-legged creature like man.

In his book *Hallmarks of Design,* engineer Dr. Stuart Burgess notes that "there are so many unique features required for bipedal motion that it is impossible for a quadruped to gradually evolve into a biped." He went on to note that evolutionists "have often claimed to have found intermediate extinct creatures between man and apes. However, in every case, the creature is either fully bipedal or fully quadrupedal, showing that it is actually either fully human or fully ape."

Today I'll mention just two of those features. First, humans have a flat face compared to apes. What does this have to do with walking? The shape of our face gives our eyes a field of view which extends all the way down to the ground in front of our feet. Being able to see the ground is important because the problem of tripping is far greater for two-legged creatures like humans.

Another structural feature is that we have strong big toes which are close to our other toes. This is important for walking and running because the final push from the ground comes from the big toe. In contrast, the big toe of apes is more like a flexible thumb that is designed for gripping branches.

Since God created us to stand, let us remember to take a strong stand for the truths found in the Word of God!

Prayer: Heavenly Father, whether I stand in Your presence or kneel before Your throne, my desire is to serve You now and forever. Amen.

Ref: S. Burgess, *Hallmarks of Design*, pp. 166-169 (Day One Publications, 2002).

Liquid Gold

Proverbs 11:28
"He that trusteth in his riches shall fall: but the righteous shall flourish as a branch."

Alchemy is considered to have played a significant role in the development of early modern science The history of alchemy is filled with fascinating stories about the precursor of chemistry and includes the names of such famous scientists as Robert Boyle and Roger Bacon. But perhaps no tale is more fascinating than the true story of glass-maker and alchemist Hennig Brand.

Name doesn't ring a bell? In the early years of the seventeenth century, Brand thought he could transform a very common substance into gold because, like gold, it was yellow in color. That substance was urine. After getting permission to collect urine from a whole camp full of German soldiers, his experiments finally paid off, but not in the way he had hoped.

Brand produced a mysterious substance that had never been seen before. The stuff glowed in the dark when exposed to oxygen, and it would also burst into flame. So he called it "cold fire" or "liquid gold."

Today the substance Brand discovered is known as phosphorus. Its name comes from the Greek word *phosphoros*, meaning "light bearing" or "luminous." Though Brand's "liquid gold" didn't make him a wealthy man, his discovery eventually led to such things as matches and phosphorus-based fertilizers.

Alchemists are still with us today. We now call them evolutionists. They are trying to transform ape-like creatures into humans. But as far as we are concerned, they would have a much better chance transforming common metals into gold!

Prayer: Heavenly Father, though riches are alluring, I know I can trust You to provide everything I need in this world ... and the next! Amen.

Ref: Richard Duncan, *Elements of Faith*, Volume 1, pp. 34-35 (Master Books, 2008).

The Living Corkscrew

Matthew 13:4
"And when he sowed, some seeds fell by the way side, and the fowls came and devoured them up:"

If you have never opened a bottle of wine, let me give you a little instruction on how to use a corkscrew. First, you puncture the cork with the corkscrew's tip. Then you turn the handle, forcing the metal screw deeper and deeper into the cork. Finally, when it is deep inside the cork, you pull on the handle of the corkscrew, removing the cork from the bottle and spilling wine all over your white shirt!

Well, what would you think if I told you that there is a seed that is shaped like a corkscrew and that it can literally plant itself in the soil! The seed of the redstem filaree is indeed a wonder to behold, especially when you see it in time-lapse photography!

As the seed dries, it changes shape and launches itself from the parent plant by a spring-type mechanism. As it falls, its corkscrew-shaped tail starts it spinning so it lands farther away from the parent plant. Once it lands, the seed pokes its head into the soil. Then its whole length starts turning in a counter-clockwise direction. With each revolution, it bores itself deeper and deeper into the soil. If the seed encounters an obstacle, it simply reverses direction, backs out and finds a better path!

All right, evolutionists, answer this question: Which came first – the seed's corkscrew shape or its "knowledge" of how to turn in the right direction until it has planted itself in the soil at the right depth? Can't answer that? Then perhaps you should ask a creationist for the answer.

Prayer: Heavenly Father, whenever I look at Your creation, I can't help but stand in amazement and praise You for Your wonderful works! Amen.

Ref: http://theawesomer.com/the-self-planting-seed/313970/.
http://en.wikipedia.org/wiki/Erodium_cicutarium. "The mechanics of explosive dispersal and self-burial in the seeds of the filaree, Erodium cicutarium (Geraniaceae)." *Journal of Experimental Biology* 214 (4): 521–529. doi:10.1242/jeb.050567.

Michael Crichton on Consensus Science

2 Corinthians 10:12
"For we dare not make ourselves of the number, or compare ourselves with some that commend themselves: but they measuring themselves by themselves, and comparing themselves among themselves, are not wise."

Darwinists and climate-change alarmists are frequently heard to say that their respective topics are a matter of "settled science." They use this term to silence critics. But is this how science is supposed to work? The late Dr. Michael Crichton, best known as the author of *Jurassic Park*, didn't think so.

In a lecture he gave at Caltech in 2003, Crichton said, "I regard consensus science as an extremely pernicious development that ought to be stopped cold in its tracks. Historically, the claim of consensus has been the first refuge of scoundrels; it is a way to avoid debate by claiming that the matter is already settled. Whenever you hear the consensus of scientists agrees on something or other, reach for your wallet, because you're being had."

Dr. Crichton wasn't done. "Let's be clear: the work of science has nothing whatever to do with consensus. Consensus is the business of politics. Science, on the contrary, requires only one investigator who happens to be right, which means that he or she has results that are verifiable by reference to the real world. In science consensus is irrelevant. What is relevant is reproducible results. The greatest scientists in history are great precisely because they broke with the consensus.... There is no such thing as consensus science. If it's consensus, it isn't science. If it's science, it isn't consensus. Period."

Creation Moments only wishes that more scientists would heed the words of evolutionist Michael Crichton. So the next time someone starts talking about "settled science," unsettle their thinking by sharing today's message with them, won't you?

Prayer: Father, through faith I know that Your Word is true, no matter how many or few believe it! Amen.

Ref: http://stephenschneider.stanford.edu/Publications/PDF_Papers/Crichton2003.pdf. "Aliens Cause Global Warming," Caltech Michelin Lecture, 1/17/03.

Darwin Doubters Must Be Punished!

Proverbs 17:26
"Also to punish the just is not good, nor to strike princes for equity."

When a scientist dares to doubt Darwin in this day and age, the science establishment goes into attack mode. It will stop at nothing to silence anyone who dares to doubt their beloved Darwin. If you think I'm exaggerating, you need to read Dr. Jerry Bergman's thought-provoking and disturbing book, *Slaughter of the Dissidents*.

In this book – the first volume in a planned three-book set – Bergman describes the experiences of several scientists who paid a high price for doubting Darwin. In the book's preface, the author summarizes the tactics used by Darwinists to punish Darwin doubters, starting with ridicule and derogatory comments.

But it gets worse, much worse. Darwin doubters are often denied admission into graduate programs. They are denied degrees. They are denied promotions or they are demoted. They will be denied tenure even if they qualify in every way. Some have been fired from their jobs. They have even received death threats.

Dr. Bergman notes that the stress associated with these various types of persecution have "often produced financial, family and other problems often ending in divorce, moving to different cities and/or states, adverse effects on children, and even suicide."

Creation Moments is also aware of many, many Darwin-doubting scientists who have not gone public with their beliefs and who live in constant fear that they will be discovered and punished. Pray for them, won't you? And pray that the science establishment would stop their hateful slaughter of the dissidents.

Prayer: Heavenly Father, I pray for those who are being persecuted by Darwinists. I pray also for their families. Most of all, I pray that Darwinists will see that what they are doing is unscientific and morally indefensible. Amen.

Ref: Jerry Bergman, Ph.D., *Slaughter of the Dissidents Volume 1,* pp. 4-6 (Leafcutter Press, 2012, Second Edition).

Life Thrives Under Antarctica's Ice

Isaiah 44:24b
"...I am the LORD that maketh all things; that stretcheth forth the heavens alone; that spreadeth abroad the earth by myself;"

Number eleven in *ScienceNews* magazine's 25 top stories of the year 2014 was the discovery of thousands of microbe varieties lurking in the rivers and lakes beneath Antarctica's massive ice sheet.

Though research teams from Russia, the United Kingdom and the United States have been looking for life under Antarctica's ice since 1999, it was the US team that eventually found and identified genetic traces of nearly four thousand microbial species or species groups.

According to a member of the US team, "The abundance of bacteria and single-celled organisms called *archaea* was a surprise." Project scientist Bruce Christner said that the number of microbes they discovered was comparable to what you would expect to find in a typical surface lake or ocean.

At this point I must say that creationists have always expected to find many signs of life under the ice sheets of Antarctica. After all, creationists know that Antarctica had a more temperate climate before the worldwide flood of Noah's time.

But to project scientist Bruce Christner, the abundance of life under Antarctica's ice is proof of something else. He thinks it bolsters the idea that life exists elsewhere in the solar system such as under Mars' polar ice caps or in a subsurface ocean on Jupiter's frozen moon Europa!

Will scientists ever stop this foolishness? When they fail to find life on Mars and Europa, will they finally admit that the Earth alone was created by God to support life? Not likely!

Prayer: Heavenly Father, I pray You will break the stony hearts of unbelievers so that they will see the Earth as Your special creation, teeming with life that You began on days five and six of Creation Week. Amen.

Ref: "Life thrives under Antarctica," *ScienceNews*, 12/27/14, p. 23.

"Evolution Is Effectively Dead"

2 Peter 3:5
"For this they willingly are ignorant of, that by the word of God the heavens were of old, and the earth standing out of the water and in the water:"

If I were to come out and say that "neo-Darwinism is effectively dead despite its persistence as textbook orthodoxy," you can be sure that evolutionists would be all over me. If I went on to say that our knowledge of genetics is now sufficient to reject evolution's slow, gradual selection of small mutational changes, you can bet evolutionists would hurl insults at me, using words I would not be permitted to repeat on this radio station.

But if one of the world's top evolutionists were to write what I just said, he would get away with it. Especially when this eminent evolutionist's name is Stephen Jay Gould. Before his death in 2002, this Harvard professor and Humanist of the Year utterly rejected slow, gradual evolution powered by mutations.

Now think about this for a moment. Gould and his esteemed colleague Niles Eldredge rejected the same neo-Darwinism that creationists reject! Both of these respected evolutionists knew that neo-Darwinism was an emperor with no clothes, so they came up with the punctuated equilibrium theory of evolution. According to Gould and Eldredge, evolution happens by way of sudden, major changes, resulting in the appearance of "hopeful monsters." Mutations produce only minor variations, said Gould. Experiments that start with flies end up with flies. So he asked: "How can such processes change a gnat or rhinoceros into something fundamentally different?"

They can't! Nevertheless, evolutionists will continue to believe in slow, gradual evolution, not because of any evidence but because their godless worldview demands it.

Prayer: Heavenly Father, I ask You to reveal Yourself to many so they will believe unto eternal life and bring glory to Your holy name! Amen.

Ref: G. Parker, *Creation: Facts of Life,* pp. 104-106. S.J. Gould, "The Return of Hopeful Monsters," *Natural History,* June/July 1977. S.J. Gould, "Is a New General Theory of Evolution Emerging?" *Paleobiology,* Winter 1980.

Tech Titans Defy Death

Psalm 90:10
"The days of our years are threescore years and ten; and if by reason of strength they be fourscore years, yet is their strength labour and sorrow; for it is soon cut off, and we fly away."

What do the billionaires who founded such tech companies as PayPal, Google, Facebook, eBay, Napster and Netscape all have in common? They are all searching for the fountain of youth, hoping to extend the human lifespan to 150 years or much longer.

As an article in the *Washington Post* points out, these billionaires are all funding projects aimed at "engineering microscopic nanobots that can fix your body from the inside out, figuring out how to reprogram the DNA you were born with, and exploring ways to digitize your brain based on the theory that your mind could live long after your body expires."

But even if it were possible to add thirty, sixty or a hundred years to our lives, would this be a good thing? According to a 2013 survey conducted by the Pew Research Center, 51 percent said they believed treatments to slow, stop or reverse aging would have a negative impact on society. And 58 percent said treatments that would allow people to live decades longer would be "fundamentally unnatural."

Well, it seems unnatural to us today that people once lived for hundreds of years. But that's what the Bible tells us. While Psalm 90:10 also informs us that we may be able to live to the age of seventy or eighty, in the end, each one of us will die. And after death, the judgment. Are you ready for that? Do you want to know that your sins have been forgiven? Then put your trust in Jesus today!

Prayer: Jesus, thank You for laying down Your life on the cross so that I can be forgiven for my many sins. By putting my trust in You for my salvation, I know I can face the future unafraid. Amen.

Ref: Ariana Eunjung Cha, "Tech titans' latest project: Defy death," *Washington Post*, 4/4/15.

"Israel Owes Us for the Ten Plagues!"

Exodus 10:4
"Else, if thou refuse to let my people go, behold, to morrow will I bring the locusts into thy coast:"

Lots of people today are skeptical that the ten plagues of Egypt described in the book of Exodus really took place. Today, however, we'll tell you about a prominent Egyptian columnist who believes that the book of Exodus is historically true even though he has no love of the Hebrew Scriptures.

Columnist Ahmad Al-Gamal recently wrote in an Egyptian daily newspaper: "We demand that the State of Israel pay compensation for the ten plagues that our forefathers in Egypt suffered thousands of years ago as a result of the curses of the Jewish forefathers." He added that Egypt should be compensated because "during forty years of wandering in the desert, the Children of Israel enjoyed our goods, which they stole before abandoning us."

The columnist's article was so outrageous, it was picked up by the Israeli press who acknowledged as historical fact everything that Al-Gamal had written. But they retorted that Egypt should first compensate Israel for keeping their Jewish forefathers in bondage for centuries and for killing all male Jewish babies just prior to the Exodus!

While today's story may bring a smile to your face, let it remind you that God is still in the business of freeing people from the bondage and penalty of sin. Through faith in Jesus Christ, you can be freed from this bondage and experience God's forgiveness. If you have not experienced such freedom for yourself, I urge you to repent and put your trust in Jesus this very minute.

Prayer: Heavenly Father, thank You for setting me free from the bondage of sin through faith in Your Son's death and resurrection! Amen.

Ref: "Egypt Demands Compensation for 10 Plagues," *Israel Today*, 4/11/14.
http://www.memri.org/report/en/print7892.htm.

Healing Power of Prayer Undeniable

__Matthew 6:33__
"Seek ye first the kingdom of God, and his righteousness; and all these things shall be added unto you."

From time to time we hear about scientific studies showing that Christians live longer and healthier lives than unbelievers. We also hear of studies proving that prayer really does work. One recent report, consisting of 1,500 different medical studies, showed that "people who are more religious and pray more have better mental and physical health."

Dr. Harold Koenig, director of Duke's Center for Spirituality, Theology and Health, said, "Studies have shown prayer can prevent people from getting sick, and when they do get sick, prayer can help them get better faster." Former atheist Tom Knox agrees: "Over the past thirty years a growing and largely unnoticed body of scientific work shows religious belief is medically, socially, and psychologically beneficial."

Though Creation Moments believes in the efficacy of prayer and the power of God to heal, we must urge caution when hearing about medical studies like these. For one thing, you can evidently believe in any "God" to enjoy the health benefits. For another, the report doesn't mention that people of faith tend to avoid risky and unhealthy behaviors. Or that the mind itself exerts a powerful force on the body.

But most important, studies like these encourage people to seek God for the wrong reasons. Good health and a long life must never be the primary motivating factors for turning to God. When you realize that Jesus voluntarily went to the cross to die for your sins, this is all the motivation you need to worship Him. Love God because He loved you first.

__Prayer: Heavenly Father, thank You for everything You bring into my life. Your presence enables me to be joyful no matter what difficulties might come my way. In Jesus' Name. Amen.__

Ref: "1,500 Medical Studies Declare Healing Power of Prayer Undeniable," ChristianHeadlines.com, 4/6/15.

Water, Water Everywhere!

John 4:10
"Jesus answered and said unto her, If thou knewest the gift of God, and who it is that saith to thee, Give me to drink; thou wouldest have asked of him, and he would have given thee living water."

NASA never seems to tire of looking for signs of life throughout the universe. Though the space agency is no longer funding SETI, the Search for Extra-Terrestrial Life, it continues searching for water because water is necessary for life as we know it.

The space agency has found oceans of water on the moons of Jupiter and Saturn. They are finding water or ice on comets, asteroids and dwarf planets like Ceres. And when they don't find water, they search for signs that water might have been present in the past.

Sadly, NASA isn't only interested in expanding mankind's store of knowledge about the universe. Though they rarely come out and admit it, their other goal is to show that the Earth is not the special place God created and filled it with life. Our planet, they believe, has life simply because water is here. And once life has begun, evolution takes care of the rest.

Chief NASA scientist Ellen Stofan says that the agency's success at finding water "inspires us to continue investigating our origins and the fascinating possibilities for other worlds, and life, in the universe." She also said, "I think we're going to have strong indications of life beyond Earth within a decade, and I think we're going to have definitive evidence within twenty to thirty years."

We will hold you to that, Ellen. In the meantime, why don't you search for the *living* water that leads to eternal life? If you seek it, you will find it for sure!

Prayer: Jesus, I pray that scientists will seek You and the Living Water that You provide because only You can satisfy the thirst for eternal life. Amen.

Ref: "The Solar System and Beyond is Awash in Water," NASA Jet Propulsion Laboratory news release, 4/7/15. "NASA Chief Scientist Ellen Stofan Predicts We'll Find Signs Of Alien Life Within 10 Years," *Huff Post Science*, 4/8/15.

Jumping Tomatoes

Exodus 8:2
"And if thou refuse to let them go, behold, I will smite all thy borders with frogs:"

"What a delicious-looking tomato," you say to yourself. But as you reach for it, that bright red tomato suddenly comes to life and jumps away in pursuit of its own lunch! That "tomato" you almost grabbed is actually a tomato frog, a creature found only on the island of Madagascar.

Just before the frog jumped out of your reach, it puffed up its shiny red body because it regarded you as a threat. Fortunately for you, the frog came nowhere near your lips. You see, when a predator bites down on a tomato frog, the frog's skin secretes a thick substance that gums up the predator's mouth and eyes – a superb defense mechanism that reminds us of the mucus secreted by hagfishes.

According to a book called *Astonishing Animals*, evolution is the reason why the tomato frog is bright red in color. The authors say it is mimicking the bright color of the poison arrow frog of Central America. But wait a minute! There *are* no poison arrow frogs on the island of Madagascar, so how could the tomato frog's bright color be attributed to mimicry? It can't mimic a frog that no creature on the island has ever seen!

Once again, when you're an evolutionist, evolution accounts for absolutely everything in nature. Creationists reject that kind of thinking. We believe that God blessed the island of Madagascar with many unique and colorful creatures. And each one was given just the right combination of characteristics by its Creator.

> *Prayer: Heavenly Father, though many people consider frogs to be creepy, I know You created them to be an essential part of our planet's ecosystem. So, creepy or not, I praise You even for the lowly frog. Amen.*

Ref: "Tomato frog," T. Flannery and P. Schouten, *Astonishing Animals*, pp. 122-123 (Atlantic Monthly Press, New York).

No Obvious Signs of Alien Life

1 John 5:12
"He that hath the Son hath life; and he that hath not the Son of God hath not life."

Not long ago Creation Moments brought you a news report about NASA's search for extraterrestrial life and how they are quite sure they'll find it by the year 2035. The primary way they're searching for life is by looking for planets and moons with water. Where there's water, they hope to find life.

But now scientists are also looking for alien life by seeking a certain kind of heat. One research team doing this is a joint project of NASA's Jet Propulsion Laboratories and Caltech, and they are basing their research on information from NASA's orbiting observatory – the Wide-field Infrared Survey Explorer.

However, after studying 100,000 galaxies, they have found no obvious signs of life in any of them.

One researcher, a professor from Penn State University, said that "the idea behind our research is that if an entire galaxy had been colonized by an advanced space-faring civilization, the energy produced by that civilization's technologies would be detectable in mid-infrared wavelengths." He likened it to the basic physics that causes your computer to radiate heat when it is turned on.

Even though no signs of alien life have been found among the 100,000 galaxies, the researchers did find fifty galaxies they plan to look at more closely. They want to see if the heat from these galaxies is coming from natural sources or from advanced civilizations.

Creation Moments could tell them the answer right now, but I'm sure they wouldn't listen. Their search for life must go on because their evolutionary worldview depends on it.

Prayer: Father, as You continue to frustrate man's search for life elsewhere in the universe, I pray that many will consider the far more important matter of how they may find eternal life in Your Son. Amen.

Ref: "Scientists find no obvious signs of life in 100,000 galaxies," *Popular Science*, 4/16/15.

Always Be Ready

1 Peter 3:15b
"...be ready always to give an answer to every man that asketh you a reason of the hope that is in you with meekness and fear:"

Not long ago, a member of our Creation Moments staff went under the surgeon's knife. Cancerous growths had been discovered through a routine colonoscopy, so to keep the cancer from spreading, the growths and his ascending colon was surgically removed.

During the pre-op office visit, the surgeon smiled and told our intrepid creationist: "Look on the bright side," he said. "In addition to removing part of your colon, I will also be removing your appendix so you'll never have to worry about appendicitis."

The patient agreed that this was, indeed, a very good thing. But his smile quickly turned to a frown when the surgeon added, "You won't miss your appendix. It's just a vestigial organ that you never really needed." What a thing to say to a creationist! Without missing a beat, our staff member asked the doctor if he was aware of studies showing that the appendix is not a vestigial left-over from our supposed ape-like ancestors. He added that the appendix is now known to be an essential part of our body's immune system.

But the surgeon wouldn't budge because this is what he had been taught in medical school. Rather than argue with the surgeon who would soon be cutting him open, our staff member decided it would be best to continue the conversation on another day. And so it shall. But he learned a valuable lesson that day. He learned that opportunities to share the truth of biblical creation can arise at any moment ... even during a doctor's appointment.

Prayer: Heavenly Father, thank You for the opportunities You give me to share biblical truth – and my faith – with others. In Jesus' Name. Amen.

Ref: This is one of the articles our staff member is sharing with his surgeon: http://www.drweil.com/drw/u/QAA401080/Do-You-Need-Your-Appendix.html.

Coat of Many Colors

Genesis 37:3
"Now Israel loved Joseph more than all his children, because he was the son of his old age: and he made him a coat of many colours."

The rainbow eucalyptus tree is undoubtedly the most beautiful tree ever. Its bark looks like it was painted by an artist using a palette full of pastel pigments. Surprisingly, this tree's coat of many colors actually comes from the way its bark peels off the tree.

According to research botanist LariAnn Garner: "As the newly exposed bark slowly ages, it changes from bright green to a darker green, then bluish to purplish, and then pink-orange. Finally, the color becomes a brownish maroon right before exfoliation occurs." She also notes that "since this process is happening in different zones of the trunk and in different stages simultaneously, the colors are varied and almost constantly changing. As a result, the tree will never have the same color pattern twice, making it like a work of living art."

The rainbow eucalyptus is also among the fastest-growing trees. It begins as a seed smaller than an ant and grows as much as eight feet per year, eventually rising up to a height of a hundred feet. And here's a surprising fact. These rainbow-colored trees are used mainly to make white paper!

Though evolutionists attribute the tree's colors to natural selection, we must point out that there are many other kinds of trees with peeling bark – including several varieties of maple trees – but the bark of these trees do not undergo a color transformation like this. Evolution had nothing to do with it. These trees are, indeed, a living piece of art painted by their Creator!

Prayer: Heavenly Father, no human artist can come close to achieving the beauty I see in Your creation! More beautiful than Your creation, however, is Your glorious plan of salvation! Amen.

Ref: L. Garner, "Under the Rainbow," *Ornamental Outlook,* 9/06. This website features many photos of the rainbow eucalyptus tree: http://www.lovethesepics.com/2013/01/living-rainbow-rainbow-eucalyptus-most-beautiful-tree-bark-on-earth-36-pics/.

Eggs-ellent Examples of Design

Job 11:18
"And thou shalt be secure, because there is hope; yea, thou shalt dig about thee, and thou shalt take thy rest in safety."

Over the years, Creation Moments has brought you countless examples of plants and animals that appear to have been designed because they really *were* designed! Nowhere is this easier to see than in the design of bird eggs.

As we mentioned on an earlier program, the shape and coloring of bird eggs are no accident. God designed them that way for a specific purpose. Today we're going to mention two other design features seen in some bird eggs.

The first are the eggs of the African jacana. These long-legged shorebirds build a flimsy nest that floats on water. When the male jacana lands in the nest to incubate the eggs, the whole nest sinks into the water. It's a good thing, then, that God created the eggs to be waterproof. This is a design feature the eggs must have had from the very beginning.

Or take the eggs of the common murre. According to *BBC Earth*, "The eggshells have cone-like structures that make the eggs 'self-cleaning'." This is useful, they say, because murre colonies are tightly packed and the eggs get showered in bird droppings. "When water lands on an egg, its water-repelling shell causes the water to gather into spherical drops" which then roll off the egg and clean it.

We could mention many other design features of bird eggs, and we will share these with you in the future. But we close today's program with praise to the God of creation who cares for all of His creatures – especially you and me!

Prayer: Lord Jesus, my heart overflows with praise when I look at Your creation and think about what You accomplished in just six days! You are awesome in every way! Amen.

Ref: "The 13 birds with the most amazing eggs," *BBC Earth*.
http://www.bbc.com/earth/story/20150319-the-birds-with-super-powered-eggs.

The Umbrella Bird

1 Thessalonians 5:5
"Ye are all the children of light, and the children of the day: we are not of the night, nor of darkness."

The best fishermen not only know where the fish are biting, but exactly what type of bait or lure is needed to catch the big ones. That's why I say that the best fisherman of all is not a man but a bird.

The black heron knows exactly where and when the fish are biting. He goes fishing for his food by wading into shallow lakes and ponds. But there's a problem. Fish avoid the water's surface to avoid the bright rays of the sun. Even if a fish does come close to the surface, the black heron is unable to see it because he is blinded by the sun's reflection.

But like I said, this bird is a master fisherman. What he does is shape his wings into a large black umbrella. He then crouches down until his wings are almost touching the water, effectively turning daylight into darkness, and attracting fish to the shade. Under cover of his umbrella, the black heron pokes his head into the water and comes out with a squirming fish in his beak.

This kind of fishing is known as canopy feeding. How did the black heron learn to fish like this? Creationists know, but evolutionists have no reasonable answer. They only have a term. Yes, evolutionists are good at coming up with terms like "canopy feeding," but when it comes to explaining how such a thing originated in the first place, they are still very much in the dark.

Prayer: Heavenly Father, I ask You to make me a good fisherman – not of fish but of men. Use me to share the gospel so that my friends and family will learn about the death and resurrection of Jesus ... and turn to Him for salvation. Amen.

Ref: "Black heron," The Website of Everything.

Problem Child

Proverbs 4:14
"Enter not into the path of the wicked, and go not in the way of evil men."

When he was just a little boy, people could tell that Charlie was a problem child. By his own admission later in life, the boy was a sadist who experienced a great deal of pleasure from torturing and killing animals. One of his favorite things to do was killing birds by pounding on their heads with a hammer.

As a teenager, Charlie continued behaving like a twisted individual. When he was seventeen, he dedicated his summer and autumns to killing animals, not for the meat but for the sheer joy of killing. When he started making plans for an ocean voyage that would take him to many exotic lands, he decided to bring along several guns, hoping to find cannibals so he could kill them.

If you haven't already guessed, I am talking about Charles Darwin. Not only was he a problem child obsessed with torturing and killing animals, he grew up to unleash a philosophy that would leave the whole world in ruins after the publication of his book *On the Origin of Species*.

As Kevin Swanson points out in his book *Apostate: The Men Who Destroyed the Christian West*, Darwin came from a long line of apostates. This included his father, Robert, and his grandfather, Erasmus, who was a pioneering humanist and enlightenment thinker. But it was the apostate Charles who turned his back on his Creator and reaches out from the grave with a philosophy that continues to destroy countless millions of lives.

> *Prayer: Heavenly Father, I pray that You will bless pastors, Bible teachers and biblical-creation ministries that are proclaiming the truth about Darwin and his godless philosophy. Show me how to be a blessing to them! Amen.*

Ref: Kevin Swanson, *Apostate: The Men Who Destroyed the Christian West*, p. 128 (Generations with Vision, 2013).

What's the Right Answer?

2 Corinthians 10:5
"Casting down imaginations, and every high thing that exalteth itself against the knowledge of God, and bringing into captivity every thought to the obedience of Christ;"

On today's Creation Moments broadcast, we're going to give you a little quiz, so put on your thinking cap!

Picture in your mind that our planet is as smooth as a billiard ball and is exactly 25,000 miles around at the equator. Now picture a 25,000-mile-long circle of rope lying on the ground that completely encircles the planet. Here's the puzzle: If you wanted to lift this unstretchable rope to a height of one foot all around the Earth, how much rope would you need to add?

Would you need to add more than ten miles of new rope? Would just a mile be enough? Less than a mile?

Most people say they would need to add more than ten miles of rope or only a mile. But hardly anyone ever gives the *correct* answer that less than a mile of rope is required. Even more surprising is that you need to add only six point two *feet* of new rope to lift the whole rope one foot high all around the planet! Though this may be hard to believe, to arrive at the correct answer, use the mathematical formula for determining the circumference of a circle – 2 Pi R.

So why does nearly everyone get it wrong? Because people are prone to say what *seems* right without thinking it through. That's why Christians, especially students, need to approach such topics as evolution with a thinking mind that chooses the *right* answer over the *popular* answer.

Prayer: Heavenly Father, I pray that You will give me a sharp mind and bold heart to stand strong against evolution no matter what the cost. In Jesus' Name. Amen.

Don't Mess with the Hoopoe!

Deuteronomy 14:12, 18
"But these are they of which ye shall not eat: the eagle, and the ossifrage, and the ospray ... And the stork, and the heron after her kind, and the lapwing, and the bat."

Don't mess with the hoopoe – especially the female! Their Creator blessed this bird with an oil gland that produces a really foul-smelling liquid. When rubbed into their plumage, it smells like rotting meat and deters not only predators but parasites as well. In fact, it also acts as an antibacterial agent.

Thanks to this nasty-smelling liquid, most predators stay far away from the hoopoe's nest. But even when a predator ignores the stench and comes looking for a meal while mama hoopoe is away, the nestlings are not defenseless. Even when they are just six days old, they can produce the same liquid and shoot it accurately into the face of predators.

With such an awful smell, perhaps it was a great blessing that God placed the hoopoe on His list of animals that were not to be eaten by His people. By the way, the King James version of the Bible refers to the hoopoe as the lapwing but we're still talking about the hoopoe. Since this bird is listed in Deuteronomy as an unclean animal, isn't it rather odd that the modern state of Israel would choose the hoopoe as their national bird?

Like the hoopoe, every person ever born – with the exception of Jesus – reeks of sin and is unclean in God's sight. This is why Jesus Christ, the sinless lamb of God, submitted Himself to die on the cross as our substitute. He will make you clean in God's sight when you repent and put your trust in Him!

Prayer: Heavenly Father, I know that there's nothing I can do to make myself clean and righteous. Thank You for sending Your Son, Jesus Christ, to die on the cross to make me acceptable in Your sight. Amen."

Ref: Wikipedia entry on "Hoopoe."

A Case Study in Scientific Fraud

Titus 1:2
"In hope of eternal life, which God, that cannot lie, promised before the world began;"

A scientific paper called "DDT: A Case Study in Scientific Fraud" was published a few years ago in the *Journal of American Physicians and Surgeons*. Though the author deals with scientific fraud related to the banning of DDT, he also gives insights into how some scientists are committing fraud to support other matters, like global warming and evolution.

The paper provides several examples of scientific fraud, but today we have time to deal with only one. Environmentalists wanted to portray DDT as a hazardous substance that was endangering birds. One scientist was able to show that the shells of bird eggs were thinner and more fragile after being exposed to the pesticide.

But here is the rest of the story. The scientist behind the thinning egg shell claim had changed the birds' diet, reducing their calcium levels from the normal 2.5 percent to just 0.56 percent. It was this radical reduction in calcium from their diet that caused the egg shells to get thinner – not the DDT! When this fact was presented to *Science* magazine, the editor refused to publish it, saying that "they would never publish anything that was not antagonistic toward DDT."

Scientific fraud – like all lies – comes from the father of lies. How different is our God. He cannot lie. And that means you can believe the Bible when it tells you that you can have eternal life when you repent and put your trust in Jesus for the forgiveness of your sins.

Prayer: Heavenly Father, I pray that scientific fraud will backfire on such deceivers. Let the lies committed in darkness get exposed to the light so they will become widely known! Amen."

Ref: J. Gordon Edwards, Ph.D , "DDT: A Case Study in Scientific Fraud," *Journal of American Physicians and Surgeons*, Volume 9, Number 3, Fall 2004, pp. 83-88.

Jumping to Conclusions

Proverbs 29:20
"Seest thou a man that is hasty in his words? there is more hope of a fool than of him."

Though they are less than a half inch in length, recent studies have shown that *Habronattus* jumping spiders have eyesight that, in some ways, rivals our own. In fact, they can do something even your family pooch can't do – see colors.

According to Nathan Morehouse of the University of Pittsburgh, "The eyes of jumping spiders could not be more different from those of butterflies or birds, and yet all three tune the color sensitivities using pigments that filter light." Scientists call this "spectral filtering," and until now, nobody suspected that a spider would be capable of it.

ScienceDaily, of course, jumps to the conclusion that the jumping spider's ability to see colors "is a remarkable example of evolutionary convergence." According to evolutionists, convergent evolution is the independent evolution of similar features in species that aren't closely related. For example, evolutionists tell us that flying insects, birds and bats evolved the ability to fly independently of one another. They are said to have "converged" on this useful trait.

This term, however, explains absolutely nothing. Evolutionists are unable to demonstrate how even one kind of creature evolved the ability to fly. Their problem is compounded when they have to explain how flight evolved three times in unrelated creatures. The real reason why these jumping spiders see color is the same reason why bats, butterflies and birds are able to fly. Because their Designer made them that way. This isn't jumping to conclusions. It is simply taking God at His word.

Prayer: Heavenly Father, thank You for giving every creature what they need to survive, and thank You most of all for what You have given to me – salvation through the shed blood of my Savior. In Jesus' Name. Amen.

Ref: Daniel B. Zurek, Thomas W. Cronin, Lisa A. Taylor, Kevin Byrne, Mara L.G. Sullivan, Nathan I. Morehouse, "Jumping spiders are masters of miniature color vision," *ScienceDaily*, 5/18/15. "Spectral filtering enables trichromatic vision in colorful jumping spiders," *Current Biology*, 2015; 25 (10): R403 DOI: 10.1016/j.cub.2015.03.033.

Half of Scientific Literature May Be Untrue

Jeremiah 9:5
"And they will deceive every one his neighbour, and will not speak the truth: they have taught their tongue to speak lies, and weary themselves to commit iniquity."

It is common for evolutionists to dismiss a lot of great work by creationists, saying that their papers do not appear in their peer-reviewed journals. But now, editors of the two most prestigious peer-reviewed medical journals have gone on record as saying that the peer-review process doesn't mean much anymore.

Several years ago, Dr. Marcia Angell, a physician and longtime editor of one of the most trusted medical journals in the world, wrote this: "It is simply no longer possible to believe much of the clinical research that is published, or to rely on the judgment of trusted physicians or authoritative medical guidelines. I take no pleasure in this conclusion, which I reached slowly and reluctantly over my two decades as an editor of *The New England Journal of Medicine*."

More recently, Dr. Richard Horton, editor of *The Lancet*, wrote the following in his own magazine: "The case against science is straightforward. Much of the scientific literature, perhaps half, may simply be untrue." He added that "science has taken a turn towards darkness" and said that this is caused by "flagrant conflicts of interest." He also said something we've been saying for years: "In their quest for telling a compelling story, scientists too often sculpt data to fit their preferred theory of the world."

So never let an evolutionist throw the peer-review process in your face as a mark of the superiority of evolution over biblical creation. Unlike peer-reviewed publications, the Bible is never wrong.

Prayer: Heavenly Father, rather than putting my trust in the fallible words of man, I put my trust in both Your written and living Word – the Lord Jesus Christ! Amen.

Ref: http://www.thelancet.com/pdfs/journals/lancet/PIIS0140-6736%2815%2960696-1.pdf. Marcia Angell, "Drug companies and doctors: A story of corruption," 1/15/09, *The New York Review of Books*.

The Old Shell Game

1 John 4:6
"We are of God: he that knoweth God heareth us; he that is not of God heareth not us. Hereby know we the spirit of truth, and the spirit of error."

Many people – especially those who believe in creation – avoid watching science shows on TV because these programs are usually saturated with unscientific claims favoring evolution. However, even Darwinian soapboxes like *Nova* can be worth watching, just as long as you do so with a discerning eye.

You especially need to pay close attention to the unwarranted assumptions and faulty logic you'll often encounter. For instance, there's "Kings of Camouflage" – a *Nova* program about the cuttlefish's amazing ability to instantaneously change the color and texture of its skin. The program did a good job of describing the color disks that expand or shrink in size.

But then the scientist added: "My guess is that their skin evolved for camouflage, because as soon as they got rid of the big shell, they had to hide from predators." Well, Mr. Scientist, why did the cuttlefish discard its shell in the first place if it was beneficial to its survival? And why didn't the cuttlefish just evolve its old shell back? Instead, evolutionists guess that the cuttlefish evolved the most complex skin of any creature on the planet as well as an amazingly complex eye and nervous system that allows it all to work.

So when you watch science shows on television, learn what you can from the true science. Enjoy the breathtaking photography as you marvel at the wondrous creatures that God has designed. And teach your children how to spot the atheistic philosophy that's camouflaged to look like true science.

Prayer: Heavenly Father, help me learn how to distinguish truth from error in a culture that knows how to make lies look so very appealing. Then help me to share the truth with others. In Jesus' Name. Amen.

Ref: http://www.pbs.org/wgbh/nova/nature/kings-of-camouflage.html.

The Sixty-Seventh Book of the Bible

Revelation 22:18
"For I testify unto every man that heareth the words of the prophecy of this book, If any man shall add unto these things, God shall add unto him the plagues that are written in this book:"

As we all know, there are 66 books in the Bible – no more, no less. And yet, from the way that many people interpret the first chapter of Genesis, it appears that they have added a sixty-seventh book to the Bible – a special "book" they use to interpret the other sixty-six books of the Bible.

The name of that sixty-seventh book is "science." Many pastors – and even entire denominations – now interpret Bible texts in a way to make them compatible with this sixty-seventh book. For example, since this "book" contends that the Earth is billions of years old, Genesis must be reinterpreted to make it "fit" into an old universe. One way they try to do this is the "day-age theory."

This theory holds that each "day" in Genesis 1 (in Hebrew *yom*) is a long period of time rather than 24 hours. It is important to note, however, that whenever *yom* appears in the Old Testament with a number attached to it, it *always* means a literal 24-hour day. In addition, whenever *yom* is modified with "evening and morning" – as it does 38 times outside of Genesis 1 – it *always* means a 24-hour day.

God's people must get back to the careful handling of God's Word. This means that we throw out the sixty-seventh book and that we start taking the Bible for what it says – not reinterpreting it in light of the latest pronouncements of so-called "science."

> ***Prayer: Heavenly Father, I know that the words of fallible men must never be used to interpret the words of an infallible God. Rather, let me interpret the world around me through the light of Your holy Word! Amen.***

Solving the Distant-Starlight Dilemma

Isaiah 45:12
"I have made the earth, and created man upon it: I, even my hands, have stretched out the heavens, and all their host have I commanded."

Scientists who reject the Bible believe that man's ability to see light from distant stars and galaxies can be likened to a deadly torpedo that sinks the ship of young earth creationism. But they are not taking into consideration that there are many possible solutions for how light from distant stars could reach the Earth very quickly.

One such theory is explained by Creation Moments Board Member Dr. Donald Clark, who holds a Ph.D. in physical biochemistry. He cites both Isaiah 42:5 and Isaiah 45:12 as evidence that the heavens are not uniform but have been "stretched out" by God. Now if the very fabric of space is stretched in parts, he explains, then light would travel much faster through these "stretchy" parts than through the non-stretched parts. He uses a guitar string as an analogy. Tightening a string produces a higher-pitched sound when plucked because it causes the sound wave to travel faster through the string. In a similar way, stretched space would cause light to travel faster through it.

In his article, "Making Sense of Light from Distant Stars," Dr. Clark writes: "Using Einstein's General Relativity equations, Mark Amunrud at the August 2013 International Conference on Creation suggested that light from the most distant galaxies ... would reach Earth in just about one week."

So while Scripture doesn't directly say that light travels much faster between the galaxies to arrive on Earth quickly, it does give us enough clues to piece together one solution to the distant-starlight dilemma.

Prayer: Heavenly Father, thank You for creating the universe in such a way that Adam and Eve didn't have to wait around for billions of years before they could gaze upon the stars You created! Amen.

Ref: Dr. Donald Clark, "Making Sense of Light from Distant Stars," 7/7/14. Creation Moments online.

What Has Evolution Given to the World?

Ecclesiastes 1:14
"I have seen all the works that are done under the sun; and, behold, all is vanity and vexation of spirit."

We get lots of interesting letters and e-mails here at Creation Moments. Most of them are positive, but occasionally we hear from someone who disagrees with us. One such writer told us that there are, and I quote, "thousands upon thousands of articles written in medical journals about how evolution has been of enormous benefit in medicine."

Want to hear how we responded? We began like this:

How many of those thousands of articles have you read? We've read some, and the references to evolution appear to have been added without thought or reason. Evolution has had absolutely nothing to do with medicine. In fact, Bible-believing surgeons in the past learned from the Bible how to wash their hands under running water. Evolution had nothing to do with the development of antiseptics. Actually, that comes from Joseph Lister, a Bible-believing scientist. Evolution had nothing to do with X-rays or Magnetic Resonance Imaging. Actually, it was a Bible-believing Christian who invented the MRI.

Evolution has played no part in the science that has given us modern technology, medicine and so much more. Evolution didn't even have anything to do with putting a man on the moon. In fact, it was a combination of applied scientists and engineers who accomplished this.

So what has evolution given mankind? Well, it has given us thousands of scientists who make their living writing about evolution and teaching it. Yes, it has created quite an industry for itself. But it has advanced science not one inch.

Prayer: Heavenly Father, I pray that You will help millions to realize that Darwin is an emperor with no clothes. Reveal to them that they have been foolish for believing that there's even a shred of truth in evolution. Amen.

Dinosaur Feathers!

Genesis 1:20
"And God said, Let the waters bring forth abundantly the moving creature that hath life, and fowl that may fly above the earth in the open firmament of heaven."

Creation Moments isn't in the business of reviewing Hollywood films, and we're not about to start now. However, we do have some thoughts about the whole Jurassic Park movie franchise. Though these films are entertaining, Christians need to be aware that the films are also promoting evolution by claiming that modern-day birds are descendents of Jurassic-age dinosaurs.

The fourth film in the franchise – *Jurassic World* – wasted no time in making this point. As the film begins, the movie screen is filled with the image of a dinosaur claw emerging from an egg, followed by a shot of the dinosaur's malevolent eye. After a sound effect of a screeching dinosaur, you see a close-up of the claw, but the camera pulls back to reveal that what you're really looking at is the claw of a modern-day bird.

Evolutionists and creationists, of course, won't fail to see what is being taught. Meanwhile, the general movie-going public receives a not-so-subtle message that the birds we see today are a product of evolution.

Many evolutionists, however, thought that the latest Jurassic Park film didn't go far enough. They were outraged that the dinosaurs didn't have feathers all over them.

How fitting that evolution is being taught in science-fiction movies! That's where it belongs! Ironically, the very existence of the Jurassic Park movies tells a story, not of chance and evolution but of design. After all, weren't the films intelligently designed by writers and directors who left absolutely nothing to chance?

Prayer: Lord Jesus, thank You for creating the universe and for creating me. Thank You most of all for entering Your creation to save me from the penalty I deserve because of my sin. As my Creator and Savior, You are worthy of all my praise! Amen.

Ancient Astronauts?

Genesis 4:21-22
"And his brother's name was Jubal: he was the father of all such as handle the harp and organ. And Zillah, she also bare Tubalcain, an instructer of every artificer in brass and iron: and the sister of Tubalcain was Naamah."

Some of you will remember the book *Chariots of the Gods* that was such a hit in the 1970s. Though its author wasn't the first or the last to write about "ancient astronauts," Erich Von Däniken popularized the notion that aliens from other worlds had come to our planet eons ago. These alien beings supposedly left traces of their visits all over the world – from the pyramids of Egypt to the Nazca lines in southern Peru.

Of course, Creation Moments does not agree with Von Däniken's conclusions. Neither does mainstream science. However, we believe that Darwinian evolution helped to make the ancient-astronauts theory somewhat credible. After all, evolutionists tell us that ancient man slowly evolved upwards. For most of mankind's history, we were supposedly little more than brute beasts, lacking both intelligence and technology. With this in mind, science was unable to explain the many out-of-place artifacts that Von Däniken wrote about in his books.

The Bible, however, tells us that mankind – going all the way back to the first man – has always been intelligent and technologically savvy. In fact, while Adam was still alive, his grandkids were already fashioning things out of brass and iron and making musical instruments.

It was the *rejection* of biblical truth that made it impossible for scientists to explain ancient technologies and out-of-place artifacts we've told you about on other Creation Moments programs. Perhaps if scientists had believed the Bible, the notion of ancient astronauts would never have taken off.

Prayer: Heavenly Father, the next time I'm asked if I believe in extraterrestrials, I will tell them I believe that Your Son came to our planet to provide salvation for Earthlings like me. In Jesus' Name. Amen.

Why Eve Was Made from Adam's Rib

Genesis 2:21-22
"And the LORD God caused a deep sleep to fall upon Adam, and he slept: and he took one of his ribs, and closed up the flesh instead thereof; And the rib, which the LORD God had taken from man, made he a woman, and brought her unto the man."

True or false: Women have one more rib than men have. You might be surprised to know how many people believe that men have one less rib because God took a rib out of Adam and used it to make Eve. The fact is that men and women have 12 pairs of ribs in our chest.

However, modern science may now have a scientific explanation on why God chose to make Eve from one of Adam's ribs.

A research team from USC has reported that humans and mice are able to re-grow removed ribs within a matter of months. The rib, in fact, is the only bone in our body that can regenerate itself. When the researchers removed rib sections and its surrounding sheath of tissue – called the "perichondrium" – the missing rib sections failed to repair itself even after nine months. However, when they removed just the rib – but left its perichondrium – the missing rib entirely repaired itself within one to two months.

As one creationist group explains it, "Many doctors know about this feature of the rib being able to regenerate itself. Quite often they will very carefully remove a rib and use it to rebuild or replace things like damaged jaw bones and eye sockets. Not only does the patient get their face reconstructed, but in time the rib grows back and they are as good as they were before the surgery."

So we see once again that the Bible makes perfect sense down to the very smallest details!

Prayer: Heavenly Father, I thank You that the Bible can be trusted from the first chapter to the last. Instill in me a greater hunger to read the Bible each and every day. Amen.

Ref: "We can regenerate! Researchers reveal our ribs regrow if damaged – and say the same could be true for our entire skeleton," *Daily Mail*, 9/16/14. "Why God Chose Adam's Rib," CreationRevolution, 10/7/11.

Will It Blend?

Deuteronomy 12:32
"What thing soever I command you, observe to do it: thou shalt not add thereto, nor diminish from it."

For several years now, the humorous "Will It Blend?" videos on YouTube have entertained millions of viewers. In each video, a man wearing a white lab coat places an object into a blender and turns it on, demolishing iPhones, Coke cans, a pair of Nikes, and anything else you can imagine. One of the only objects he wasn't able to "blend" was a crowbar. He didn't even try.

Though the videos might crack you up, they remind Creation Moments of something that's really quite serious. Many people throughout history have tried to blend Christianity with other religions to make it more acceptable to people of other faiths. But all they end up doing is stripping Christianity of its power, its truth and its Savior.

Webster's dictionary defines "syncretism" as "the combination of different forms of belief or practice." From the earliest days of Christianity, syncretism has been a blight on orthodox, biblical Christianity. It has taken many forms and has been successful at deceiving millions. Even today there are people who are trying to blend the Bible with evolution – as in theistic evolution. But like that crowbar, the Bible and evolution just will not blend ... so don't even try.

Without knowing it, however, many Christians have "blended" their beliefs with the belief systems of non-Christians. That's why it is so important for each one of us to examine the things we believe as well as the traditions we practice – placing them not in a blender but under the "microscope" of the Word of God.

Prayer: Heavenly Father, I pray that Your Holy Spirit will make me aware of any beliefs in my heart that are not based on scriptural truth so that I may turn away from them in repentance. In Jesus' Name. Amen.

The Death Knell of Christianity?

Psalm 2:4
"He that sitteth in the heavens shall laugh: the Lord shall have them in derision."

"I think that evolution is absolutely the death knell of Christianity." No, those aren't my words. These words were spoken by atheist Frank Zindler, and they make it very clear what the origins debate is really all about. The issue of origins is nothing less than the front line in the battle between two incompatible faiths.

In step-by-step fashion, Zindler described how he reached that conclusion. "The most devastating thing that biology did to Christianity," he said, "was the discovery of biological evolution." In other words, he starts out by declaring that evolution has been proven, which, as you know, is very far from the truth.

He goes on to say that evolution means that Adam and Eve never really existed. In his own words, "Now that we know that Adam and Eve never were real people, the central myth of Christianity is destroyed."

And why is that? He continues, and I quote: "If there never was an Adam and Eve, there never was an original sin. If there never was an original sin, there is no need of salvation. If there is no need of salvation, there is no need of a savior. And I submit that puts Jesus, historical or otherwise, into the ranks of the unemployed."

Yes, atheists foolishly think that evolution is the death knell of Christianity. In reality, biology – with its established fact that life comes only from life – points directly to our Creator. Biology is actually the death knell of atheism and evolution.

> *Prayer: Lord Jesus, I pray that You will use me to help others see that life comes only from life and that the original Life who started it all was You! Amen.*

Ref: Frank Zindler quote taken from debate with William Lane Craig held in 1993 at Willow Creek Community Church.

Four Eyes Or Two?

Job 5:8-9
"I would seek unto God, and unto God would I commit my cause: Which doeth great things and unsearchable; marvellous things without number:"

Behold the amazing *cuatro ojos* – the four-eyed fish! At first glance, you would say it has just two eyes. But look closer and you'll see that each eye is divided into two separate eyes!

According to the book *1000 Wonders of Nature*, "Light from the world above the water passes through the width of the egg-shaped lens, which gives very good long-distance vision. While swimming just below the water's surface, the fish can simultaneously check the world above for predators, such as birds, while viewing the surface and depths below for food."

Evolutionists at one website point out that the fish's "two large, bulbous eyes are positioned on the upper side of the head" and are "split horizontally into two sections by a band of tissue. The upper lobe is flattened while the lower lobe is rounded, allowing it to see clearly both above and below water." The lens of the eye also changes in thickness from top to bottom to deal with the different refractive indices of air and water.

Despite the obvious design features of these eyes, evolutionists can't see it because they are blinded by their atheistic worldview. Without evidence of any sort, they say that the fish's eyes "evolved specifically for the purpose of exploiting the narrow ecological niche between aquatic and terrestrial habitats."

Evolutionists are completely in the dark when it comes to explaining how such a primitive creature as the *cuatro ojos* came to have such complex eyes. Creationists are not in the dark about it because we accept that those eyes were created that way.

Prayer: Heavenly Father, thank You for giving me eyes to see Your marvelous creation and for giving me the faith to know it was all created by Your Son, Jesus Christ. Amen.

Ref: "How Two Eyes Serve as Four," *1000 Wonders of Nature*, p. 198 (Readers Digest). Animal-World website on four-eyed fish.

Jet-Propelled Squid

Psalm 8:8-9
"The fowl of the air, and the fish of the sea, and whatsoever passeth through the paths of the seas. O LORD our Lord, how excellent is thy name in all the earth!"

You've heard of flying squirrels and flying fish, but did you know there's also a squid that can fly? Even more surprising is that it accomplishes this amazing feat using jet-propulsion! Yes, though they were once only rumors, we now know that a certain kind of squid can emerge from the water and streak through the air at speeds up to 25 miles per hour. A team of Japanese marine biologists have actually photographed Japanese flying squid doing just that in groups.

Jun Yamamoto at Japan's Hokkaido University describes how they take flight. The squid normally swims backwards through the water using its fins. But after taking in a large quantity of water, the squid forcefully ejects the water through a nozzle near its head and becomes airborne.

"Once they finish shooting out the water," Yamamoto said, "they glide by spreading out their fins and arms. The fins and the web between the arms create aerodynamic lift and keep the squid stable on its flight arc. As they land back in the water," he said, "the fins are all folded back into place to minimize the impact."

Of course, it didn't take long for evolutionists to claim that these squid evolved all the parts necessary to fly so that they could escape from sea-going predators. But as it turns out, Japanese flying squid face even more formidable predators in flight – like the albatross. So once again we see that the evolutionary explanation just doesn't fly.

Prayer: Lord Jesus, the wonders of Your creation have turned many people away from evolution. Use me to share creation truth with my family and friends so I can help them find the path that leads to You. Amen.

Ref: "Japanese flying squids – Scientists confirm flight capabilities," Digital Journal/Science, 2/10/13.

The Five-Second Rule Scientifically Tested

Leviticus 11:33
"And every earthen vessel, whereinto any of them falleth, whatsoever is in it shall be unclean; and ye shall break it."

Do you suppose that bacteria talk to one another when food is dropped on the floor nearby? Their leader shouts: "Hey, don't eat that! We have to wait five seconds!"

Apparently, that's what many of us must think. According to the well-known five-second rule, it is safe to eat food that has fallen to the floor if we pick it up in less than five seconds. If you follow this rule, you're in good company. Research shows that 87 percent of people said they would eat food dropped on the floor or that they've already done so.

Professor Anthony Hilton, from Aston University's School of Life and Health Sciences, has now shown that there's some truth to the five-second rule. Food picked up just a few seconds after being dropped really is likely to have less bacteria on it than if it is left on the floor for a longer period of time. Bacteria are also less likely to contaminate food dropped on carpeted rather than hard surfaces. And bacteria are much faster at jumping aboard moist than dry food.

However, Professor Hilton warns, "Consuming food dropped on the floor still carries an infection risk as it very much depends on which bacteria are present on the floor at the time." He adds, however, that the findings of this study "will bring some light relief to those who have been employing the five-second rule for years, despite a general consensus that it is purely a myth."

Prayer: Heavenly Father, thank You for giving my body a great deal of protection against bacteria. But I thank You most of all for protecting me from the temptations that the world, the flesh and the devil throw at me. In Jesus' Name. Amen.

Ref: Aston University, "Dropped your toast? Five-second food rule exists, new research suggests," *ScienceDaily*. 3/10/14.

Hallucigenia

Psalm 113:3
"From the rising of the sun unto the going down of the same the LORD'S name is to be praised."

When a very bizarre animal was first discovered in the Burgess shale of the Canadian Rocky Mountains, scientists literally couldn't tell its head from its tail. In fact, when they reconstructed the worm-like creature, they got it totally wrong. They assembled it upside down and put its head where its tail should be.

Even after finally getting it all sorted out, the creature still looks like something you'd see in a hallucination or nightmare. That's why they called it *Hallucigenia*. This odd-looking animal was a little over two inches long and had seven pairs of nail-like spines sticking out from its back. On its underside were an equal number of flimsy legs, tipped with claws. And it had a single row of tentacles hanging down from its neck.

Using sophisticated imaging technology, scientists learned that the head was actually at the end of a long, tube-like neck. Near the end of the head were two bean-shaped eyes. "Below the eyes, like an almighty grin, sits a ring of teeth," said University of Cambridge paleontologist Martin Smith.

With all these various body parts, this is pretty complex for an animal without an ancestor, wouldn't you say? Yes, the Burgess shale and the Cambrian explosion have always been a nightmare for evolutionists. After all, how can such a wide variety of highly complex creatures suddenly pop into existence out of nowhere? If they evolved, what did they evolve *from?* A good question, and we're *still* waiting for a credible answer from evolutionists!

Prayer: Heavenly Father, though You filled our world with many bizarre creatures, I thank You for each one of them because they serve as a testimony to Your great creative genius! Amen.

Ref: "Hallucination? No, this Cambrian period creature was really weird," Digital Journal | Science, 6/24/15.

Bubonic Plague, a Problem for Evolutionists

Psalm 38:7
"For my loins are filled with a loathsome disease: and there is no soundness in my flesh."

Want to ruffle the feathers of an evolutionist? Just remind him that mutations are harmful. Yes, there are many mutations that are neutral, but beneficial mutations are only a figment of the evolutionist's imagination. The truth is that mutations typically bring about deformities and death. Such is the case with the tiny mutation that changed a harmless organism found in the intestines of humans and animals into a bacteria that resulted in the bubonic plague.

I am talking about the genetic alteration of the *Yersinia pestis* bacterium. Research conducted at the Northwestern University Feinberg School of Medicine has discovered that a single, tiny genetic change fundamentally altered this bacterium into a killer that took the lives of more than 75 million people in 14th century Europe.

In this case, the mutation caused what is known as the "black death." In other cases, mutations cause cancer and a host of other deadly diseases. Think about it. Why do x-ray technicians cover your body with lead shielding when you get x-rays if the radiation might bring about a beneficial mutation?

As this new research makes clear, mutations are harmful and can even bring about the death of millions. Neither do mutations lead to new species. If anything, they lead to the *extinction* of species. Despite the evidence, evolutionists persist in claiming that mutations plus natural selection over millions of years add up to every form of life on our planet. However, the lack of beneficial mutations is the kiss of death to evolution.

> **Prayer: Father, when the first Adam disobeyed You, suffering and death were unleashed upon our world. I thank You that Your Son – the Last Adam – brought to our world healing and forgiveness. Amen.**

Ref: "Tiny genetic shift led to 'The Black Death' and worse," Digital Journal | Science. "Early emergence of *Yersinia pestis* as a severe respiratory pathogen," Nature Communications, 6/30/15.

How the Seahorse Got Its Square Tail

Genesis 1:31a
"And God saw every thing that he had made, and, behold, it was very good."

Evolutionists and creationists alike enjoy gazing upon the elegant seahorse swimming lazily about the aquarium. However, only creationists are able to see that the seahorse bears the unmistakable stamp of design – all the way from the top of its horse-shaped head to the tip of its unique square tail.
Yes, I said square tail. This striking design makes the seahorse considerably better at gripping and grasping than if it had curved sections in its tail. When you look closely, you will see that the tail is made up of about 36 square-like segments that progressively taper off in size along the length of the tail. By using a 3D printer to duplicate the tail and submitting it to crushing tests, an international team of researchers has now discovered that the squared sections also provide the seahorse with much better armor than a cylindrical tail.
"Almost all animal tails have circular or oval cross-sections – but not the seahorse's," said Michael Porter, lead investigator and assistant professor in mechanical engineering at Clemson University. Porter's research group is now using 3D printing to help them develop new structures and robotic systems.
Although evolutionists will tell you that the seahorse's unique tail only *appears* to be designed, they act as if the tail really *was* designed! After all, how intelligent would it be for them to pursue new technologies based on the tail of a seahorse that has no intelligence behind it at all?

> **Prayer: Heavenly Father, it is so obvious to me that creatures like the seahorse were designed by You. I pray that You will use me to make others aware that all of creation points directly to You! In Jesus' Name. Amen.**

Ref: "Why the seahorse's tail is square," Phys.org, 7/2/15.

What's This? A Warm-Blooded Fish?

Psalm 8:8-9
"The fowl of the air, and the fish of the sea, and whatsoever passeth through the paths of the seas. O LORD our Lord, how excellent is thy name in all the earth!"

Though the fish story I'm about to tell you is absolutely true, it goes against what we've always been taught about fish being cold-blooded. Scientists have just discovered the very first warm-blooded fish. Though fish typically require heat from the environment to stay warm, the opah is able to elevate its own temperature without the help of its environment.

So how does the opah pull off this trick? Biologist Nick Wegner tells us, "The opah appears to produce the majority of its heat by constantly flapping its pectoral fins, which are used in continuous swimming." But the opah's *real* secret was discovered when scientists examined the opah's unique gills.

As an article at *LiveScience* explains, "The blood vessels are set up so that the vessels carrying cool, oxygenated blood from the gills to the body are in contact with the vessels carrying warm, deoxygenated blood from the body to the gills. As a result, the outgoing blood warms up the incoming blood, a process called counter-current heat exchange."

Hey, wait a minute! Doesn't this sound like the "wonder net" heat-exchange system in birds that we described in a Creation Moments program called "Why birds don't need socks"?

Evolutionists will have to say that this heat-exchange system evolved not only in birds but in fish as well. But we have a much better explanation. The same God who created birds on the fifth day of creation used the same design on some birds, which He also created on day five.

Prayer: *Heavenly Father, it makes perfect sense to me that You would invent a heat-exchange system and then use it for more than one kind of creature. Similarities of design indicate that all of creation had one Creator! Amen.*

Ref: "First Warm-Blooded Fish Found," *LiveScience*, 5/14/15.

Carbon-14 Being Found in Dinosaur Fossils!

Psalm 25:5
"Lead me in thy truth, and teach me: for thou art the God of my salvation; on thee do I wait all the day."

Evolutionists were so sure that dinosaur fossils are too old to contain any carbon-14, they never even bothered to check. Or perhaps they were afraid of what they would find. In any case, creationist scientists are now boldly going where evolutionists have feared to tread.

An article in the spring 2015 edition of the *Creation Research Society Quarterly* presents never-before-seen carbon dates for 14 different fossils, including a Triceratops and other dinosaurs. As ICR scientist Brian Thomas points out, "Because radiocarbon decays relatively quickly, fossils that are even 100,000 years old should have virtually no radiocarbon left in them. But they do."

Using the services of five different commercial and academic laboratories, the research team tested seven dinosaur bones and detected carbon-14 in them all. The team also looked at several dozen published carbon-14 results for fossils, wood and coal throughout the geologic column and found radiocarbon in almost 50 samples.

Evolutionists, of course, try to explain it all away by saying the samples were contaminated. Nice try, but if this were true, how could it be said that *any* radiocarbon dating results are accurate? Evolutionists can't just toss out data they don't like!

Or maybe they can. In this way, they are acting like police officers telling bystanders, "Nothing to see here, folks. Just move along." If anything, this type of behavior strongly suggests that something *really is* going on! And in this case, it's proof positive that dinosaurs are thousands and not millions of years old!

Prayer: Heavenly Father, thank You for raising up creationists who are uncovering evidence that the Earth is young. I pray that this will cause many people to abandon their belief in evolution. Amen.

Ref: Brian Thomas, M.S., "Carbon-14 Found in Dinosaur Fossils," 7/6/15.

Scientists Still Don't Have a Clue How Life Began

John 6:29
"Jesus answered and said unto them, This is the work of God, that ye believe on him whom he hath sent."

In a typical anti-creationist op-ed piece for *Scientific American*, columnist John Horgan begins by explaining why he titled his column: "Pssst! Don't tell the creationists, but scientists don't have a clue how life began."

Horgan looks at an Arizona State University status report on research into the origin of life. In Horgan's own words, the scientists concluded, "Geologists, chemists, astronomers and biologists are as stumped as ever by the riddle of life."

He also notes that he was most startled by the revelation that more scientists than ever are believing in the directed panspermia theory. As we have mentioned on previous Creation Moments broadcasts, this is the notion proposed by DNA co-discoverer Francis Crick that aliens came to Earth in a spaceship and planted the seeds of life here billions of years ago. I guess that's like saying that ET didn't phone home after all but started a family of his own here on Earth.

Though he admits that this notion fails to explain how those aliens came into existence, Horgan quickly claims that creationists have the same problem in explaining how life first began. After all, he writes, "What created the divine Creator?"

Of course, this is a question that was answered centuries ago. By definition, God is the uncreated Creator. Of course, columnist Horgan won't agree with this answer. But he should stop misleading his readers by claiming that creationists are just as perplexed about the origin of life as evolutionists are.

Prayer: Heavenly Father, evolutionists would rather believe in the most far-fetched theories rather than believe in You. I pray that You will change some of them from doubters into believers! In Jesus' Name. Amen.

Ref: Brian Thomas, M.S., "Pssst! Don't tell the creationists, but scientists don't have a clue how life began," *Scientific American*, 2/28/11.

Giant Redheaded Centipede

Genesis 8:19
"Every beast, every creeping thing, and every fowl, and whatsoever creepeth upon the earth, after their kinds, went forth out of the ark."

Many of us would enjoy watching a redheaded woodpecker bang bang bang his head into a tree trunk. But today I'd like to tell you about another redheaded creature that would probably have many of us running in the opposite direction. I'm talking about the giant redheaded centipede.

After the Texas Parks and Wildlife Department posted a picture of the creepy crawler on its social media pages, the photo quickly went viral. To see one in real life, you would have to travel to northern Mexico or the southern United States. But we have a better idea. Check out the photo at the Creation Moments website.

As their name makes abundantly clear, these centipedes are really big – their bodies are typically eight inches long. And with 21 to 23 pairs of long yellow legs, they are able to get around quickly. Those legs aren't just for locomotion, by the way. They can actually inject venom after making tiny incisions in human skin. Plus, the centipede has a pair of fangs that can inject a painful toxin. While it has not been confirmed, there are reports of one person dying of a heart attack after being bitten.

Though the giant redheaded centipede might send shivers up our spine, it reminds us that we live in a fallen world. As the Bible tells us, we long for a better country, a heavenly home where there is no pain, no suffering, and ... I sincerely hope ... no giant redheaded centipedes!

Prayer: Heavenly Father, Adam's disobedience brought suffering and death into the world, but the Last Adam endured suffering and death so that we may be reconciled to You. Thank You for sending Your Son to rescue us! Amen.

Ref: Elizabeth Palermo, "Giant Redheaded Centipede Photo Goes Viral, Horrifies the Internet," *LiveScience*, 7/10/15.

Evolutionists – Masters of Mimicry

Ephesians 5:6
"Let no man deceive you with vain words: for because of these things cometh the wrath of God upon the children of disobedience."

While doing some research on how the non-venomous hognose snake is able to mimic the behaviors of the deadly rattlesnake, we happened to come across a website with an interesting article on biological mimicry.

According to the website, "Biological mimicry occurs when a group of organisms – the mimics – have evolved to share common perceived characteristics with another group – the models – through the selective action of a signal-receiver – or dupe." In this instance, the hognose snake is the mimic, the rattlesnake is the model, and the creature that missed out on having the hognose snake for lunch is the dupe.

They go on to say that mimicry in most cases is advantageous to the mimic and harmful to the dupe. With this in mind, I would like you to consider the proposition that evolutionists themselves are "mimics." They mimic the "model" scientists who deal with real science – that is, those involved in empirical, observable and repeatable science.

Meanwhile, these mimics in white lab coats get paid handsome salaries and receive government grant money so they can enjoy all the advantages of appearing to be scientists engaged in real science. At the same time, however, they are harming other people – the dupes – by getting them to believe a lie. Worse yet, they make it easy for the dupes to abandon belief in God.

In a world that fails to distinguish between real science and a godless philosophy masquerading as science, let us all be diligent in teaching others how to tell the two apart.

Prayer: Heavenly Father, I pray that You will continue to give Your people wisdom and discernment so that we won't be fooled into exchanging the truth for a lie. Amen.

Ref: "Biological Mimicry," The Encyclopedia of Earth website, 9/24/09.

This Plant Calls Out to Bats!

Leviticus 11:19
"And the stork, the heron after her kind, and the lapwing, and the bat."

Imagine this – you're a carnivorous pitcher plant with no scent to attract insects. Even worse, though most pitcher plants are designed to kill insects in their pool of digestive fluids, you don't have enough fluid to do the job. What's a poor pitcher plant to do?

Well, God didn't forget about the needs of the *Nepenthes hemsleyana*. He gave it a unique ability to "call out" to bats that are looking for lodging. Once the bat and its family are checked in for the night, they pay for their room by depositing enough nitrogen-rich guano to keep the plant healthy and alive.

So how exactly does a plant without vocal cords call out to bats? The plant's leaves are actually shaped in a way that reflects sonar. As bats fly through the jungles of Borneo – using echolocation to get around – their "clicks" are reflected to them by the leaves of this plant. They take this as an open invitation to bed down for the night.

If you ever look inside one of these plants, you won't find insects in various stages of decomposition. Instead, you'll find the Hardwicke's woolly bat roosting comfortably inside, either alone or with her whole family.

Here at Creation Moments, we never tire of telling others about the amazing plants and animals God has created. Judging from the fact that we've been doing this for over 50 years, many of you feel the exact same way. So tell your friends about Creation Moments, won't you?

> ***Prayer: Heavenly Father, though I cannot see You, I enjoy looking upon the works of Your hands! I pray that You will give me many opportunities to share the gospel and the truth of biblical creation with others. In Jesus' Name. Amen.***

Ref: Schöner, Schöner, Simon, Grafe, Puechmaille, Ji & Kerth. 2015. "Bats Are Acoustically Attracted to Mutualistic Carnivorous Plants," Current Biology. http://dx.doi.org/10.1016/j.cub.2015.05.054.

Glow, Little Cockroach

Job 29:3
"When his candle shined upon my head, and when by his light I walked through darkness;"

As Creation Moments has reported in the past, evolutionists were dead wrong when they told us that viceroy butterflies evolved to look like toxic monarch butterflies to trick predators into thinking that they, too, are toxic. As it turned out, viceroy butterflies really *are* toxic – even more toxic than monarchs!

Despite the facts, the myth of viceroy butterfly evolution was taught to millions of unsuspecting children. And it worked so well, evolutionists are at it again. According to *National Geographic News*, the rare *L. luckae* cockroach evolved the ability to glow in the dark "to mimic the bioluminescent click beetle, whose glow warns predators of its toxicity."

The only evidence they've got for this is that the wavelengths of light released from both bugs are identical. That's it? That's the best evidence they've got?

The *National Geographic News* article admits that the cockroach's bioluminescence also gives it an evolutionary *disadvantage* because it makes it easier for predators to spot them. But evolutionary biologist Olivia Judson already has an answer for that: "Bioluminescence is like any evolutionary tool – there is no single use for it. It can attract, deter, or even be used as an invisibility cloak of sorts."

It's been said that when all you have is a hammer, everything looks like a nail. In the same way, when all you have is a godless natural world, everything – including a bioluminescent cockroach – looks like it can be used to trick people into believing that evolution caused it.

Prayer: Lord Jesus, I pray that You will shine Your light on the hearts and minds of evolutionists so that they will turn away from evolution and stop deceiving so many. Amen.

Ref: N. Mott, "Glowing Cockroach Mimics Toxic Beetle," *National Geographic News*, 8/30/12.

A Snake with Four Legs?

Genesis 3:14
"And the LORD God said unto the serpent, Because thou hast done this, thou art cursed above all cattle, and above every beast of the field; upon thy belly shalt thou go, and dust shalt thou eat all the days of thy life:"

Did you hear the exciting news about the scientist who found a snake with four legs? Well, actually it wasn't a snake. It was a fossil that British paleontologist David Martill from the University of Portsmouth discovered in an almost-forgotten box at a museum in Germany.

After the fossil's discoverer announced: "I realized we'd actually got the missing link between lizards and snakes," it didn't take long for tabloid-style headlines like these to start appearing:

- "Four-legged fossil holds secret of snake's slithering origins"
- "Weird four-legged snake was transitional creature"
- "Snakes' four-legged missing link"

Headlines like these are very misleading, to say the least. For one thing, this fossil is not the earth-shaking discovery that science publications would like us to believe. Have you ever heard of worm lizards? Did you know that many snakes with two legs have been found? The only thing different about this fossil is that it has four rather than two legs. Some scientists, in fact, aren't even convinced it's a snake at all. They think it's just a reptile.

Here at Creation Moments, we think it's quite possible that this four-legged snake fossil is hard evidence for what creationists have been saying all along about snakes. For as the Bible tells us in Genesis 3:14, God punished the serpent for deceiving Eve by making it crawl on its belly. This strongly suggests that snakes had legs before the curse.

Prayer: Heavenly Father, give Your people a calm spirit so that we will not panic when scientists discover a fossil that appears to confirm evolution. I am confident that no fossil will ever prove Your Word to be untrue! Amen.

Ref: "Four-legged snake fossil found," *Science Daily*, 7/23/15.

Hawking Joins Search for Extraterrestrial Life

Jeremiah 29:13
"And ye shall seek me, and find me, when ye shall search for me with all your heart."

In an interview with Britain's *The Guardian* newspaper in 2011, famed physicist Stephen Hawking warned that the efforts by the SETI Institute to actively "contact possible intelligent life out there is way too risky and dangerous."

In a report on why he fears contact with alien life, Hawking wrote: "I imagine they might exist in massive ships, having used all the resources from their home planet. Such advanced aliens would perhaps become nomads, looking to conquer and colonize whatever planets they can reach. If aliens ever visit us, I think the outcome would be much as when Christopher Columbus first landed in America, which didn't turn out very well for the Native Americans."

Fast forward now to July of 2015, and you'll see that professor Hawking has become much more hawkish in searching for extraterrestrial life. Stephen Hawking joined tech investor Yuri Milner and famed astrophysicist Lord Martin Rees in announcing the launch of Breakthrough Initiatives. Over the next 10 years, Milner will invest $100 million in this project that will give scientists access to some of the best telescopes in the world. This, they feel, is one giant step forward in the search for extraterrestrial life.

However, they will come up just as empty-handed as the SETI project did after decades of fruitless searching. Breakthrough Initiatives is searching for the wrong thing, you see. If they searched for God – the giver of all life – they would find Him when they searched with all their heart.

Prayer: Father, I know that the search for aliens is really just an attempt to prove that our planet is nothing special. I pray that many scientists will turn to You after failing in this endeavor to find life elsewhere. Amen.

Ref: Dave Masko, "UFO sightings warn of alien invasion, states Stephen Hawking who fears SETI's search," HULIQ, 5/28/11. Heather Kelly, "Search for alien life gets $100M and Stephen Hawking's blessing," CNNMoney, 7/21/15.

Spider Performs Amazing Engineering Feat

Proverbs 30:28
"The spider taketh hold with her hands, and is in kings' palaces."

Have you ever stood at a construction site watching cranes lift heavy objects suspended by steel cables? Well, a spider in Madagascar can do the same thing. They routinely lift snail shells twenty times their weight using stronger-than-steel cables made of spider silk. This would be like a 180-pound piano mover lifting a piano weighing 3,600 pounds!

Olios coenobitus is an elusive spider that lives in snail shells which it hoists high up in bushes to protect itself from predators. Though the snail shell spider was first discovered in 1926, it wasn't until 2011 that a BBC camera crew was able to film one as it raised a snail shell off the jungle floor using a network of silk threads.

The spider begins by attaching a silk thread to the branch of a bush and then sticking the other end on the shell. He then attaches another silken strand but makes it a bit shorter than the first. Each time he makes the thread shorter and shorter. After doing this multiple times, the shell lifts off the ground and then heads upwards, where it will eventually become a roomy shelter for the spider.

With a brain no bigger than a grain of rice, this spider is able to pull off one of the most amazing engineering feats ever! How did that spider survive before it started hoisting shells many times its weight? And who taught it such complex engineering skills? The answer is obvious. But evolutionists stubbornly refuse to admit it.

Prayer: Heavenly Father, You left signs of Your creative ingenuity all over Your creation. I praise You that even a lowly spider can give powerful testimony to an unbelieving world. In Jesus' Name. Amen.

Ref: Ella Davides, "Madagascar's elusive shell-squatting spider filmed," BBC Earth News, 2/8/11.

Science Was Wrong About Boa Constrictors

Genesis 3:13
"And the LORD God said unto the woman, What is this that thou hast done? And the woman said, The serpent beguiled me, and I did eat."

Remember being taught that evolution gave boa constrictors the ability to kill their prey by literally squeezing the breath out of them? Well, scientists now know this isn't true at all.

As a recent *National Geographic* article points out, "Boa constrictors were long thought to kill their prey by suffocation, slowly squeezing the life out one ragged breath at a time. But a new study reveals that these big, non-venomous serpents, found in tropical Central and South America, subdue their quarry with a much quicker method: Cutting off their blood supply."

The study was conducted by vertebrate ecologist Scott Boback and his research team at Dickinson College in Carlisle, Pennsylvania. After feeding lab rats to boa constrictors, they found that the snakes were killing their prey by stopping the flow of blood – a much faster killing method than death by strangulation. Boback theorizes that killing by circulatory arrest gives constricting snakes an evolutionary advantage. After all, the quicker the prey dies, the less chance that it will bite the snake.

That's what it's all about, isn't it? Evolution. Isn't it interesting that evolution was also given credit when science mistakenly thought that suffocation was the means of death? The important thing to evolutionary scientists is not *how* boa constrictors kill their prey. The important thing is giving evolution full credit.

When creationists observe the animal kingdom, we praise God. When evolutionists observe the animal kingdom, they, too, give praise, but to a different "god" entirely.

Prayer: Heavenly Father, it seems that evolution has become a god to many. I pray that You will expose this false god and point people to the Creator who alone can save people from their sins. In Jesus' Name. Amen.

Ref: Jason Bittel, "Why We Were Totally Wrong About How Boa Constrictors Kill," *National Geographic*, 7/22/15.

Charles Darwin and Karl Marx

Psalm 53:1
"The fool hath said in his heart, There is no God. Corrupt are they, and have done abominable iniquity: there is none that doeth good."

Have you noticed how many of the most admired atheists come from a science background? I could name dozens – Richard Dawkins, Sam Harris, Daniel Dennett and Stephen Hawking, to name a few. Thousands of people are fans of the late Carl Sagan and his protégé, Neil deGrasse Tyson. Some of these atheists are practically superstars with followers numbering in the millions.

Well, Charles Darwin had his share of followers, too. At first, his *Origin of Species* was accepted not on any scientific merit but rather because it offered an apparently rational alternative to the miraculous. Darwin's earliest followers were not scientists of the day but rather theologians who rejected the miracles of the Bible.

Darwin's followers each had their own particular motives for accepting the theory of evolution. But one of his most notable followers was Karl Marx. As the author of *The Communist Manifesto*, Marx found the struggle-to-the-death principle in natural selection a perfect confirmation of his own view of man's class struggle. In appreciation, Marx sent Darwin a copy of his *Das Kapital* in 1873.

Six years later, Marx wrote to Darwin, requesting permission to dedicate his next volume to him. But Darwin declined the offer, explaining that "it would pain certain members of his family if he were associated with so atheistic a book."

Sometimes I wonder if Darwin would turn in his grave if he knew of the horrors his theory has been asked to justify in the last two centuries!

Prayer: Heavenly Father, I do not follow any mortal man. Instead, I follow Your Son, Jesus Christ, who died to give me eternal life. He alone is worthy of my praise. Amen.

Ref: Ian T. Taylor, *In the Minds of Men: Darwin and the New World Order,* pp. 386-387 (TFE Publishing and Creation Moments, Sixth Edition, 2008).

Comfort from Biblical Creation

Psalm 31:7
"I will be glad and rejoice in thy mercy: for thou hast considered my trouble; thou hast known my soul in adversities;"

I hope that today's program doesn't sound like a two-minute-long infomercial for Creation Moments' *Letting God Create Your Day* books. The purpose of today's broadcast is to point out that creation truth is not only a good way to begin conversations with your friends about spiritual things. It is also good for your own emotional health.

Let me explain. One of our regular listeners recently wrote to tell us that he was feeling depressed by what he was reading in the daily newspaper. So he bought one of our *Letting God Create Your Day* books to read something different.

He told us that his appreciation of God grew as he read story after story about God's loving care for even the lowliest of creatures. He explained that if God provided for the needs of ants, sparrows, jellyfish and even amoebae, he knew that this same God would take care of his needs as well. By the time he finished the book, his depression was completely gone.

The truth of biblical creation isn't just good for sharing your faith with others. Let God comfort you through the Creation Moments broadcast as we look more closely at His amazing creation. As you consider how God cares for each one of His creatures, remember that God won't be any less faithful in providing for the needs of those who have been redeemed by the precious blood of His Son, Jesus Christ.

Prayer: Heavenly Father, forgive me when I take my eyes off Your Son and allow anxiety to rule in my heart. As this sinful world continues on its downward spiral, I know I can find lasting peace only by looking unto Jesus. Amen.

Ref: Each volume in the *Letting God Create Your Day* series contains hundreds of scripts from our international radio program. They are published by Creation Moments and are available online at www.creationmoments.com or by calling 1-800-422-4253 during regular office hours.

Mr. Peanut

Proverbs 3:5-6
"Trust in the LORD with all thine heart; and lean not unto thine own understanding. In all thy ways acknowledge him, and he shall direct thy paths."

Born the son of slaves in the early 1860s, George Washington Carver grew up to become one of America's most prominent chemists. Though most people recognize him as the man who gave us peanut butter, few know that Carver was a scientist with a very close relationship with his Creator.

As we learn from the book, *George Washington Carver: His Life and Faith in His Own Words* by William Federer, after speaking to the U.S. House and Ways Committee, the committee chairman asked him where he had learned so much about peanuts. Carver responded: "From an old book." The chairman asked: "What book?" And Carver replied, "The Bible."

Wishing to know more, the chairman then asked, "Does the Bible tell about peanuts?" And Carver responded, "No sir. But it tells about the God who made the peanut. I asked Him to show me what to do with the peanut, and He did."

On a different occasion, when asked the secret of his success, he answered: "It is simple. It is found in the Bible, 'In all thy ways acknowledge Him and He shall direct thy paths.'"

While working in his laboratory – which he called "God's Little Workshop" – Carver developed over 300 innovative uses for peanuts and another 100 uses for the soybean, including beverages, cosmetics, paints, medicines, and food products.

So the next time an evolutionist tells you that men of faith can't be real scientists, tell him the story of Mr. Peanut.

Prayer: Heavenly Father, You take delight in humble men and women who acknowledge You as Creator. I pray that You will continue to raise up scientists of faith so they will glorify Your Son, Jesus Christ. Amen.

Ref: Federer, William J. (2002). *George Washington Carver: His Life and Faith in His Own Words*. St. Louis: Amerisearch.

Fastest Tongue in the West

Genesis 7:14
"They, and every beast after his kind, and all the cattle after their kind, and every creeping thing that creepeth upon the earth after his kind, and every fowl after his kind, every bird of every sort."

Considering the chameleon's slow walking speed, you might never guess that it possesses the fastest tongue of any creature. On second thought, it *does* makes a lot of sense for a wise Designer to equip chameleons with a tongue that's actually longer than its entire body ... and lightning-fast so it can catch fast-moving insects. Using high-speed video and x-ray film, two Dutch biologists have calculated that the chameleon shoots its tongue out of its mouth at more than 26 body lengths per second. They found that the chameleon's tongue accelerates from 0 to 20 feet per second in only 20 milliseconds!

So what's the chameleon's secret? By dissecting the chameleon's tongue, the biologists discovered a layer of collagen tissue that serves as a biological catapult that propels the tongue similar to the way a drawn bow fires off an arrow.

Can evolution explain how this catapult got there? Creationist and author Bruce Malone points out in his book *Censored Science*, "It is theoretically possible to make up a story of how the chameleon's specialized eyes, tongue, and catapult structures all just happened to develop independently in order to work in perfect concert. But given that there is no fossil evidence of any other creatures possessing such structures, it would seem to be a more logical conclusion that the features were created all at the same time, rather than developed by many small step-changes over huge periods of time."

Prayer: Father, though evolution makes sense to some, when people actually look closely at the creatures You created, evolution makes no sense whatsoever. Thank You for telling us the true story in the Bible! Amen.

Ref: B. Trivedi, "'Catapults' Give Chameleon Tongues Superspeed, Study Says," *National Geographic News*, 5/19/04. Bruce Malone, *Censored Science: The Suppressed Evidence,* pp. 18-19 (Search for the Truth Publications, 2009).

Peer-Review Fraud Strikes Again and Again

Proverbs 20:17
"Bread of deceit is sweet to a man; but afterwards his mouth shall be filled with gravel."

Evolutionists never tire of taunting creationists with the statement, "If creationism is scientific, why are there no papers from creationists published in peer-reviewed publications?"

The standard creationist response is that since these peer-reviewed publications are run by evolutionists, it is obvious why they refuse to publish articles by creationists. But perhaps we should also mention that when an article appears in a peer-reviewed publication, this is no guarantee that the articles are accurate or even true.

While the peer-review process makes sense in theory, in actual practice, it has been found to fail miserably in this fallen world of ours. Not long ago, we told you about two of the most prestigious medical publications admitting that the peer-review process is deeply flawed. One editor wrote: "It is simply no longer possible to believe much of the clinical research that is published." The editor of the other publication added: "Much of the scientific literature, perhaps half, may simply be untrue."

Now, several new cases of peer-review fraud have come to light. One of the biggest research publishers in the world recently announced that 64 articles published in 10 of its scientific journals are being retracted because the scientists had come up with a scheme to peer review their *own* papers! To make matters worse, the publisher retracted 43 other papers five months earlier for the same reason.

Because the sin nature dwells in all men, we know that the only writings that can be fully trusted were authored by the ultimate truth-teller – God Himself.

Prayer: Heavenly Father, I know that it is impossible for You to lie. How blessed are all those who put their trust in You and Your holy Scriptures! In Jesus' Name. Amen.

Ref: B. Yirka, "Publisher retracts 64 articles for fake peer reviews," Phys.org, 8/19/15. "Half of Scientific Literature May Be Untrue," Creation Moments online.

The Importance of a Worldwide Flood

2 Peter 2:5
"And spared not the old world, but saved Noah the eighth person, a preacher of righteousness, bringing in the flood upon the world of the ungodly;"

A long-time friend of Creation Moments, Richard Rothermel, recently shared his thoughts with us about the worldwide flood of Noah's time and why this historic event is so important for biblical creationists to defend. Here is what he had to say.

"The secular world cannot deal with the relics of the Global Flood, as described in Scripture. Instead, unbelievers prefer to interpret the sedimentary rocks and fossils as the result of vast ages of geological and biological evolution. Old-earth creationists also deny a global flood, preferring the interpretations of secular science (and inadvertently, perhaps, damaging the foundation of the Gospel)."

He continues: "The 'secular' interpretation is not only anti-biblical, it simply cannot stand scientific scrutiny. As biblical creationists, we must make every effort to ensure we understand the significance of the Global Flood for upholding the authority and integrity of Scripture. The Global Flood truly 'washes away' all those millions of years, and Darwinism goes SPLAT!

"An enormous mass of geological and paleontological information has come to light over the years which confirms the reality and magnitude of the Flood. With all of this evidence, Dr. John Whitcomb, coauthor of the groundbreaking *The Genesis Flood,* has rightly observed: 'Now the Christian world has no excuse – if they ever had any – for adding millions and billions of years to earth history.' Until we master the concepts of the Global Flood chronology and its proper interpretation," Richard concludes, "we will have a hard time supporting biblical six-day creation and refuting compromise."

Prayer: Heavenly Father, I pray that more people will come to see that the evidence for a worldwide flood is overwhelming. In Jesus' Name. Amen.

Ref: Personal correspondence from Richard Rothermel on file at Creation Moments. Used with permission.

Planned Parenthood's Roots in Darwinism

Proverbs 20:17
"Bread of deceit is sweet to a man; but afterwards his mouth shall be filled with gravel."

"Planned Parenthood, the leading government-funded abortion provider in the United States, has a history rooted in Social Darwinism, a sociological theory from evolution's doctrine of survival of the fittest."

With these words, Creation Moments' Mark Cadwallader comments on the underlying reasons on why Planned Parenthood has now been shown to commit sickening atrocities the likes of which have not been seen since the systematic slaughter of millions by the Nazi war machine.

"Evolution," he continues, "regards humans as animals – just more evolved. And as Darwin's theory came into acceptance, so-called experts began treating human beings as glorified animals. Videos have come to light showing doctors calmly discussing the manner in which they crush and dismember babies to preserve decidedly human body parts and organs for sale. Hopefully, such exposés will awaken many people to a mass human atrocity in the same way that images of the holocaust did."

He goes on to write that "an original objective of Planned Parenthood has been to cleanse society of undesirable people – including minorities – so that evolution could be advanced." He then points out that such evil acts are "intellectually justified by the fallacy of evolution."

Christians are feeling more and more like exiles in a world that has grown increasingly hostile to biblical foundations. Just as Daniel and his three friends refused to compromise their faith in the time of their exile, so must the church in our day do what we can to expose and oppose the rotted fruit of evolutionary ideas.

Prayer: Father, I pray that You will bring Planned Parenthood and other abortion providers to utter ruin. Until that day, I pray You will change the minds of many who now feel that abortion is their only option. Amen.

Ref: Mark Cadwallader, "Has Compromise with Evolution Led to a Babylonian Exile for the Church?" Creation Moments online.

How Tears Point to a Designer

Psalm 6:6
"I am weary with my groaning; all the night make I my bed to swim; I water my couch with my tears."

Many men today believe that the shedding of tears is a sign of weakness. They seem to have forgotten that Jesus wept when told that Lazarus had died. Jeremiah is known as the "weeping prophet" who shed tears over a backslidden nation. Even David, the slayer of tens of thousands, wrote that he often drenched his bed with tears.

Our Designer not only gave us glands to produce tears, He also gave us an ingenious way to dispose of excess tears. Here is how William Paley described this design feature in his nineteenth-century classic *Natural Theology*:

"To keep the eye moist and clean ... a wash is constantly supplied by a secretion for the purpose; and the superfluous brine is conveyed to the nose through a perforation in the bone as large as a goose quill. When the fluid has entered the nose, it spreads itself upon the inside of the nostril and is evaporated by the current of warm air which in the course of respiration is continually passing over it."

Listen closely to Paley's next words: "It's easily perceived that the eye must want moisture; but could the 'want' of the eye generate the gland which produces the tear, or bore the hole by which it's discharged – a hole through bone?"

The answer, of course, is no, not in a billion years! As any rational person can plainly see, these features were designed by our Creator.

Prayer: Heavenly Father, thank You for supplying my eyes with tears so that I can see more clearly. Thank You also for giving me the eyes of faith to see that I can have eternal life by trusting in the completed work of Your Son on the cross. It's in His Name I pray. Amen.

Ref: William Paley, *Natural Theology,* p. 33 (The Works of William Paley, D.D., Ward Lock & Co., London).

Meet a Creationist Who's a Nuclear Physicist

Titus 2:1-2
"But speak thou the things which become sound doctrine: that the aged men be sober, grave, temperate, sound in faith, in charity, in patience."

Evolutionists never seem to tire of taunting creationists with such words as: "If creationists were *real* scientists, their papers would be published in secular scientific journals."

But Dr. Brandon van der Ventel, professor of physics at Stellenbosch University in South Africa, doesn't allow such statements to go unchallenged. "Creation scientists," he said, "have published many articles in secular journals." He adds, however, that papers from many creation scientists "are ignored and rejected" not because they are faulty, but because they reject "the evolutionary (or billions-of-years) paradigm."

Though this professor once believed that "somehow God must have used evolution," he now rejects that. "If the story of the fall of Man is 'mythology'," he says, "then there is no need for a plan of salvation. This is ultimately an attack on the personage of our Lord Jesus Christ and His redeeming work on the cross."

Though most of his fellow academics reject a worldwide flood, he says that "Jesus explicitly refers to Noah's flood.... This raises the question: If the Lord's statement about Noah's Flood is false, then why should we believe His statement concerning eternal life?"

Dr. van der Ventel turned away from evolution when he realized that the Bible is authoritative even on scientific matters. Doubting Genesis, he said, often leads to "a slippery slide to unbelief, where we can end up questioning the inerrancy and infallibility of the entire Bible."

Prayer: Heavenly Father, I pray that many other scientists in academia will turn to You in faith and that they will inspire others to turn away from evolution. Amen.

Ref: J. Sarfati, "Nuclear physicist embraces biblical creation," Creation Ministries International.

The Fish Fountain

1 Corinthians 10:13
"There hath no temptation taken you but such as is common to man: but God is faithful, who will not suffer you to be tempted above that ye are able; but will with the temptation also make a way to escape, that ye may be able to bear it."

One of my favorite books is Rupert Sheldrake's *Dogs That Know When Their Owners Are Coming Home*. Though the book was written by a non-Christian author who is a well-qualified scientist from Cambridge, this book is filled with interesting stories about animal abilities that should leave evolutionists speechless.

For instance, the author describes a school of fish as resembling a large organism. To paraphrase from the book, the school's members swim in tight formations, wheeling and reversing in near unison. When the school turns to the right or the left, individuals formerly on the flank assume the lead. When under attack, a school may respond by leaving a gaping hole around a predator. More often, the school splits in half and the two halves turn outward, swim around the predator, reverse direction, and eventually rejoin each other. This "fountain effect" leaves the predator ahead of the school. Each time the predator turns, the same thing happens.

Some filmmakers have produced fine video-footage taken in tropical waters, showing schools of fish, perhaps several hundred, all swimming in very close formation. Something alerts them and they all turn in exactly the same direction in a micro-second; there is no confusion, no touching.

Clearly, there is one superb Mind in absolute control of each of those unfettered souls to guide and to keep them and to teach us. That Mind, of course, is the one who created the fish and the oceans in which they live. Jesus is His name.

Prayer: Father, thank You for providing me with a way to escape temptation and for sending Your Son to provide the only way to escape eternal torment. Amen.

Ref: R. Sheldrake, *Dogs That Know When Their Owners Are Coming Home and Other Unexplained Powers of Animals*, p. 158.

Darwin's Headhunters

Genesis 1:27
"So God created man in his own image, in the image of God created he him; male and female created he them."

Charles Darwin introduced his theory of evolution in 1859 in the book *Origin of Species*. Many evolutionists, however, would like people to forget that the book's full title is *On the Origin of Species By Means of Natural Selection or the Preservation of Favoured Races in the Struggle for Life*.

It is undeniable that Darwin promoted racist ideas that could then be solidly founded upon "science." Although he never mentioned Man in his *Origin of Species*, the implications were there that Man had come from the animal kingdom.

In his third book, *The Expression of Emotions in Man and Animal*, published in 1872, Darwin left no doubt about his racist ideas, and head-measuring became a full-time occupation to determine which races were more "evolved" than others. Teams of hunters set out to obtain specimen heads from aboriginal Australian and Tasmanian owners ... whether the owners had ceased to use them or not! Entire tribes were shot down, and the museum cases of the "civilized" nations became filled with neatly graduated rows of heads giving supposed "proof" of evolution.

Most evolutionists deny this gruesome chapter in the history of Darwinism. Yet it shows that Darwinism is not only racist, it treats humans as if they were animals. Is it any wonder, then, that most evolutionists are in favor of abortion and euthanasia while creationists oppose these practices? After all, creationists – not evolutionists – know that all people, young and old, are created in the image of God.

Prayer: Heavenly Father, I look forward to the day You put an end to the evil slaughter of unborn babies, each of whom bears Your image. Amen.

Ref: Ian Taylor, "Scientific Racism," on file at Creation Moments.

Immigration Problem

Exodus 22:21
"Thou shalt neither vex a stranger, nor oppress him: for ye were strangers in the land of Egypt."

On an earlier broadcast, I shared with you how Darwinism led to racism and horrendous acts that are comparable to what terrorists are doing today with unspeakable acts of violence. Though our broadcast today tells a far less gruesome story, it does show how Darwinism even played a part in America's early immigration policies.

At the turn of the twentieth century, when immigrants were flooding into North America, there were certain groups very hostile to Roman Catholicism, and they were looking for justification to exclude Catholics from the country on scientific grounds. Obviously, to have excluded them on religious grounds would have undermined the boast of America being "the land of the free" and would in any case have violated the Constitution.

However, Franz Boas, a German anthropologist, was asked to set up a team on Ellis Island where immigrants could be scientifically screened. Screening was carried out by measurement of head size and distance from fingertip to kneecap, all on the presumption that Man had evolved from the ape and that those closer to the ape's proportions were to be excluded. The "dark races" especially included those from Catholic countries. Many immigrants were turned back on the dubious grounds that their lower intelligence would dilute the general gene pool.

Even today there are people who believe that certain races of people are less evolved than others. But those people certainly did not get that notion from the Bible. The Bible affirms there is only one race – the *human* race!

Prayer: Heavenly Father, like the song says, Jesus loves the little children, red, brown, yellow, black and white. Thank You, Jesus, for coming to save the lost, no matter what their skin color might be. Amen.

Ref: Ian Taylor, "Scientific Racism," on file at Creation Moments.

No Brain? No Problem!

Psalm 136:25-26
"Who giveth food to all flesh: for his mercy endureth for ever. O give thanks unto the God of heaven: for his mercy endureth for ever."

Judging from the things many of us choose to eat, you'd be right in thinking that we sometimes don't use our brains when it comes to making healthy food choices. But a tiny multicellular animal, which scientists call "Trix" for short, doesn't use its brain at all at mealtime. In fact, it doesn't have a brain or a single nerve cell in its whole millimeter-sized body.

So how is it that these tiny, brainless creatures are found worldwide, crawling across shallow seafloors on a belly covered in hair-like cilia in search of algae? Scientists have recently discovered that Trix have a whole bag of tricks that enables them to seek out algae with surprising sophistication.

According to an article in *LiveScience*, when scientists used light and electron microscopy to look up close at Trix, they found two previously unknown cell types, giving Trix a total of six body cell types in all. By contrast, humans have hundreds of different cell types. Scientists think that one of these new cell types – which they call "crystal cells" – allows Trix to sense its environment, thus enabling it to find its lunch on the seafloor.

Though scientists now know how Trix find food, they don't have a clue how they came to have these crystal cells to begin with. And they are utterly incapable of explaining how Trix survived before they supposedly "evolved" those cells. To creationists, however, it's crystal clear that God created this creature with all the cell types it needed to survive.

Prayer: Heavenly Father, it doesn't take a brain like Einstein's to see that even an animal without a brain is highly complex. I praise Your Son, Jesus Christ, for being that Designer! Amen.

Ref: N. Weiler, "No Organs, No Problem: Weird Animal Hunts Without Nerves or Muscles," *LiveScience*, 9/2/15.

Beautiful But Deadly Woman

Esther 1:11
"To bring Vashti the queen before the king with the crown royal, to shew the people and the princes her beauty: for she was fair to look on."

In Italian, its name means beautiful woman. The belladonna plant first received this name because it was used to make eye-drops for women in ancient Rome. Belladonna made the women appear more seductive by dilating the pupils of their eyes.

Today, the plant is often called the deadly nightshade, and it certainly lives up to that name. The deadly nightshade is one of the most poisonous plants known. It is particularly dangerous because its glistening black berries are tempting to children. But they can also be fatal to an adult who foolishly consumes as few as two of the berries. Ingesting even one leaf can kill an adult, but the roots are the deadliest part of all.

The deadly nightshade is one of a group of hallucinogenic plants that are known to produce a sensation of flight. According to the book *Secrets of the Natural World*, "Witches were thought to have used [the] plants in their potions, which may explain why witches are typically portrayed flying through the night sky."

Creation Moments believes that the deadly nightshade represents a collaboration of sorts between God and Adam. God originally created the plant with berries that could be eaten without harm. After all, God called His creation "very good." But after Adam took that fateful bite of the forbidden fruit, God cursed the ground. This is when we believe that the chemistry of the harmless belladonna was changed into a plant that can kill with one bite.

Prayer: Heavenly Father, thank You for telling us in the Bible how death and suffering came into the world so that we can know they were not a part of Your original creation. In Jesus' Name. Amen.

Ref: *Secrets of the Natural World,* p. 122 (Reader's Digest/Dorling Kindersley, 1993).

NASA's Glory Has Departed

1 Samuel 4:21
"And she named the child Ichabod, saying, The glory is departed from Israel: because the ark of God was taken, and because of her father in law and her husband."

Some of you listening to me today are too young to remember the glory days of NASA – when the U.S. space agency put a man into Earth orbit and, a few years later, on the surface of the moon. NASA had brains and guts in those days – as evidenced by the remarkable story of Apollo 13 ... or when astronauts weren't afraid to read passages from Genesis and give thanks to the God of the Bible in their transmissions from space.

That is changing, though. One of NASA's main missions now is helping Muslims feel better about their scientific accomplishments. NASA is also trying to find life on other planets to prove that evolution can happen anywhere. And the space agency is spending millions of taxpayer dollars to prove that man-made global warming is real.

So how are these missions working out? Well, a few years ago a NASA rocket, carrying a satellite called Glory, was launched, only to plummet a few minutes later into the Pacific Ocean. How sad, considering that its mission was to gather data supporting man-made global warming!

According to the Associated Press, the Glory represented "the second-straight blow to NASA's weakened environmental monitoring program. The same thing happened to another climate-monitoring satellite [two years earlier] with the same type of rocket." A NASA official noted that "the loss of Glory will mostly hurt projections and modeling of future climate change."

Truth be told, the glory departed long ago when NASA became focused on missions supporting such pseudosciences as man-made global warming and evolution.

Prayer: Heavenly Father, thank You for scientists who are involved in real scientific endeavors that benefit mankind. In Jesus' Name. Amen.

Ref: Steve Schwartz, "NASA – The Glory Has Departed," 3/9/11. Creation Moments online.

World's Top Intellectuals Are Theists

1 Kings 4:30
"And Solomon's wisdom excelled the wisdom of all the children of the east country, and all the wisdom of Egypt."

"Have you ever heard the claim 'all smart people are atheists', or maybe its inverse: 'people who believe in God are dumb'?" So begins a column at Examiner.com by a writer who decided to see if this is true by taking a closer look at the ten highest IQs ever recorded. His conclusion is that the most intelligent people on the planet believe in God.

Of course, we don't have time today to tell you about all ten, but we will give you a glimpse into the amazing mind of Andrew Magdy Kamal. Just before he turned 17, Andrew was found to have the highest IQ ever recorded – 231.734. According to the columnist, Andrew "is a staunch conservative and a member of the Republican Tea Party. He is also the founder of the Coptic Orthodox Messianicans Group." Andrew said he "hopes to use his talents and intelligence to spread the news of Messiah Yeshua (Jesus Christ) his hero."

The highest IQ ever recorded for an adult in the "advanced IQ test" was attained by Abdessellam Jelloul. He scored an adult IQ of 198. When asked about his religious beliefs, he replied that he "believes in God, a Supreme Architect of the universe."

Obviously, there are intelligent people who do and who don't believe in God. In the end, it is important to remember that it is not one's intellect that can put a person into a right relationship with the God of the universe. It is faith in Jesus Christ.

> ***Prayer: Heavenly Father, I thank You that it is not a person's great intellect that causes a person to turn to Your Son for salvation. Otherwise, I might not be praying to You right now! Amen.***

Ref: S. Williams, "Of 10 highest IQ's on earth, at least 8 are Theists, at least 6 are Christians," Examiner.com, 7/10/14.

Eyewitness Account of the Exodus?

Exodus 10:22
"And Moses stretched forth his hand toward heaven; and there was a thick darkness in all the land of Egypt three days:"

In the early 19th century, a papyrus was found in Egypt – called the Ipuwer papyrus – that appears to be an actual eyewitness account of the events recorded in the book of Exodus. Of course, nearly all secular archaeologists attempted to say the papyrus does not describe these events. After all, their worldview is opposed to the historic reliability of the Bible.

But Rabbi Mordechai Becher, senior lecturer for the Gateways Organization, points out on his webpage that "the papyrus accurately describes violent upheavals in Egypt, starvation, drought, escape of slaves (with the wealth of the Egyptians), and death throughout the land." He goes on to write, "The papyrus was written by an Egyptian named Ipuwer and appears to be an eyewitness account of the effects of the Exodus plagues from the perspective of an average Egyptian."

Here are just a few eyewitness observations from the Ipuwer papyrus that match the events in the book of Exodus: plague is throughout the land, the river (Nile) becomes blood and men thirst for water, grain has perished on every side, the hearts of animals weep while cattle moan, the land is without light, the children of princes have died, and precious metals and stones are "fastened on the neck of female slaves."

In all, Rabbi Becher points out 18 similarities where the Ipuwer papyrus lines up with the Scriptures. For us, the evidence couldn't be more convincing. The events described in the book of Exodus – like the events recorded in Genesis – are historically accurate.

Prayer: Heavenly Father, though I need no archaeological evidences to "prove" that the Bible is true, I thank You for these evidences because they may help convince skeptics to take a closer look at the Bible. Amen.

Ref: Rabbi Mordechai Becher, "The Ten Plagues Live from Egypt."

The Best Camouflage of All

Genesis 1:28a
"And God blessed them, and God said unto them, Be fruitful, and multiply, and replenish the earth, and subdue it: and have dominion over the fish of the sea..."

If you held a *salpa maggiore* in one hand and looked at it closely, what you would see is ... your hand! That's because this strange marine invertebrate is transparent! Put it back in the water, and it would almost completely disappear. A fisherman who found one floating near the surface of the water said, "It felt scaly and was quite firm, almost jelly like, and you couldn't see anything inside aside from this orange little blob inside it."

Paul Cox, director of conservation and communication at the National Marine Aquarium, said: "The salp is barrel-shaped and moves by contracting, pumping water through its gelatinous body. It strains the water through its internal feeding filters, feeding on phytoplankton from the upper sunlit layer of the ocean. They have an interesting life-cycle with alternate generations existing as solitary individuals or groups forming long chains."

Getting back to its transparency, Cox observed, "In common with other defenseless animals that occupy open water ... the translucence presumably provides some protection from predation. Being see-through is a pretty good camouflage in water."

Indeed it is. And the *salpa maggiore* couldn't have acquired such camouflage through mutations no matter how many billions of years you allow for evolution to take place. In the hands of an all-powerful Creator, however, such remarkable camouflage is no problem at all. In fact, God specializes in creating animals and plants that shut the mouths of evolutionists and open the mouths of His followers with words of praise!

Prayer: Now unto the King eternal, immortal, invisible, the only wise God, be honour and glory for ever and ever. Amen.

Ref: Leon Watson, "Now that's a jelly fish! Stunned fisherman catches wobbly shrimp-like creature that's completely see-through," DailyMail.com, 1/21/14.

Science Makes the Case for God

Romans 1:20
"For the invisible things of him from the creation of the world are clearly seen, being understood by the things that are made, even his eternal power and Godhead; so that they are without excuse."

Though Creation Moments doesn't agree with everything that Christian author and apologist Eric Metaxas believes about origins, we have to congratulate him on his excellent op-ed piece that appeared in *The Wall Street Journal* – "Science Increasingly Makes the Case for God."

He pointed out, "Today there are more than 200 known parameters necessary for a planet to support life – every single one of which must be perfectly met, or the whole thing falls apart."

He followed that up by asking such pertinent questions as: "Can every one of those many parameters have been perfect by accident?" And: "Doesn't assuming that an intelligence created these perfect conditions require far less faith than believing that a life-sustaining Earth just happened to beat the inconceivable odds to come into being?"

Naturally, an op-ed piece like this was sure to anger atheists and evolutionists. One prominent atheist – Lawrence Krauss – wrote a letter to the editor stating what we've heard many times before: "The appearance of design of life on Earth is overwhelming, but we now understand, thanks to Charles Darwin, that the *appearance* of design is not the same as design."

Following in Richard Dawkins' footsteps, Krauss discards what he himself calls the "overwhelming" evidence of design. But atheists like Krauss and Dawkins must toss out this evidence only because their godless worldview demands it. Creationists prefer *real* science over the atheistic *philosophy* of evolutionism. Things that appear to be designed really *were* designed!

Prayer: *Father, thank You for creating the universe with such complexity that even atheists can't miss seeing Your hand in its creation. In Jesus' Name. Amen.*

Ref: Eric Metaxas, "Science Increasingly Makes the Case for God," *The Wall Street Journal*, 12/26/14, p. A11.

Where Did the Elephant Get Its Trunk?

Genesis 1:24
"And God said, Let the earth bring forth the living creature after his kind, cattle, and creeping thing, and beast of the earth after his kind: and it was so."

On a previous Creation Moments program, we told you about the elephant's trunk with its 40,000 muscles. This is 70 times the number of muscles in your entire body! And that amazing trunk can bulldoze a tree or pick up a pin.

But why does an elephant have a trunk in the first place? For the answer to that question, we looked in a very old book called *Natural Theology*. It was written by William Paley in 1802. Yes, that's the same William Paley who came up with the famous analogy of a person who finds a watch and deduces that the watch must have had a watchmaker.

According to Paley, "The short unbending neck of the elephant is compensated by the length and flexibility of his proboscis," or trunk. "He could not have reached the ground without it." In other words, the elephant needed such a trunk to reach food and water. But, then, a person could ask why does the elephant have such a short and unbending neck? Paley answers that he needed such a neck to support such a heavy head!

Obviously, the elephant was given its trunk by a very wise Creator. As Paley so well pointed out, if the elephant had waited millions of years for his trunk to grow, "how was the animal to subsist in the meantime... until the prolongation of its snout was completed?" It is a question that evolutionists today are still unable to answer.

Prayer: Heavenly Father, thank You that even the elephant's trunk provides a witness to Your hand in creation! In Jesus' Name. Amen.

Ref: William Paley, DD, *Natural Theology,* 1809, p. 140 (*The Works of William Paley,* Ward Lock & Co.).

Bullies!

1 John 4:6
"We are of God: he that knoweth God heareth us; he that is not of God heareth not us. Hereby know we the spirit of truth, and the spirit of error."

In a recent study of over 4,000 children in the United Kingdom and more than 1,420 children in the United States, researchers concluded that bullied children have similar or worse mental health problems later in life than those who have suffered emotional or even physical abuse.

Yes, bullying is a very serious issue, and it's not just other kids who are doing the bullying. Sometimes it's a teacher who pounds his desk and shouts, "I've been a science teacher for over thirty years, and I tell you that evolution is not a hypothesis or even a theory – it's a fact!" Most students are reluctant to say anything for fear of receiving a failing grade. Besides, their teachers have studied such matters far longer than they have, so what right do they have to say anything?

But here's another way of looking at this. By rejecting the Bible as the authoritative source of knowledge, such teachers have a distorted understanding of the world around them. And the longer they look through this distorted "lens," the more hardened they become in their error.

So, Christian students, don't throw out your beliefs when you are bullied by arrogant science teachers. Remember that their long years of study proves nothing. After all, the Bible – unlike science textbooks – doesn't need to be constantly updated and revised. The Bible's truths are true for all time. So the longer you study the Bible – the reliable source of knowledge and wisdom – the greater your understanding of the world around you will be.

Prayer: Heavenly Father, with evolution being taught as fact, I pray that You will place Your hand of protection on Christian students and help them stand strong for biblical truth. In Jesus' Name. Amen.

Ref: "Back off, bullies!" *ScienceNews for Students*, 5/12/15. Steve Schwartz, "We're Not Worthy! We're Not Worthy!" 10/26/09, Creation Moments online.

Ancient Birds Flew Over Dinosaurs' Heads

Genesis 1:21
"And God created great whales, and every living creature that moveth, which the waters brought forth abundantly, after their kind, and every winged fowl after his kind: and God saw that it was good."

As the T-Rex approaches its prey, the dinosaur is suddenly hit from above with something white that spatters on its upturned head. Yes, it's the same kind of bird droppings you find all over your car when you park under the branches of a tree.

What's that again? Evolutionists have been telling us that modern-day birds evolved from dinosaurs. The two kinds of animals weren't around at the same time, they say. But now, a report tells us that one "ancient bird's intricate arrangement of the muscles and ligaments controlled the main feathers of its wings, supporting the notion that at least some of the most ancient birds performed aerodynamic feats in a fashion similar to those of many living birds."

Dr. Luis M. Chiappe, the investigation's senior scientist, said, "The anatomical match between the muscle network preserved in the fossil and those that characterize the wings of living birds strongly indicates that some of the earliest birds were capable of aerodynamic prowess like many present-day birds."

Of course, the research team didn't want people to end up thinking that birds haven't evolved over the past 125 million years, so they mentioned that the bones of the fossilized bird is "skeletally quite different from their modern counterparts." In addition, the press release was accompanied with an artist's rendition of the bird. To preserve the evolutionary story, the drawing shows a bird with an open beak filled with sharp teeth.

Prayer: Lord, birds and dinosaurs were created one day apart so they lived at the same time. Thank You for the scientific evidence confirming that the Bible is true. Amen.

Ref: "Tiny ancient fossil from Spain shows birds flew over the heads of dinosaurs," *ScienceDaily*, 10/6/15, Natural History Museum of Los Angeles County.

Now You See It, Now You Don't!

Psalm 104:24
"O LORD, how manifold are thy works! in wisdom hast thou made them all: the earth is full of thy riches."

What tiny shrimp-like creature is a bright iridescent blue one moment and then – in the blink of an eye – invisible the next? It's the *Sapphirina* copepod, also known as the "sea sapphire." Though the sea sapphire is only about a tenth of an inch long, it is surely one of the most amazing creatures in God's creation.

As an article at the Deep Sea News website points out, "The secret to the sea sapphire's shine is in microscopic layers of crystal plates inside their cells. In the case of blue sea sapphires, these crystal layers are separated by only about four ten-thousandths of a millimeter, about the same distance as the wavelength of blue light."

Now here's where the creature's vanishing act comes into play. "When blue light bounces off these crystal layers, it is perfectly preserved and reflected. But for other colors of light, these small differences in distance interfere, causing the colors to cancel out."

And what happens when they cancel out? The creature vanishes! You really need to see this. If you go to the Creation Moments website and search on the term "sea sapphire," you can then copy a link in the footnotes and paste it into your web browser to see a video of the sea sapphire in action.

The sea sapphire's vanishing act surpasses what even the most accomplished illusionists can do. Its unique abilities clearly point to a Creator who never ceases to amaze and delight us with His magnificent creatures!

Prayer: Heavenly Father, only You could have given Your creatures so many amazing characteristics. It's silly to think for even one moment that these marvelous creatures are the product of evolution. Amen.

Ref: R.R. Helm, "The Most Beautiful Animal You've Never Seen," Deep Sea News.

Smithsonian Scientist Scolds National Geographic Society

Psalm 7:11
"God judgeth the righteous, and God is angry with the wicked every day."

Here at Creation Moments, we frequently get upset with the National Geographic Society for misleading their readers and viewers with stories supporting evolution. But when a scientist at the Smithsonian Institution writes a scorching letter to them to express his outrage, this is something that doesn't happen every day!

Let me quote from that letter, written in 1999 by Storrs Olson, who, at the time, was Curator of Birds at the Smithsonian's National Museum of Natural History.

Olson's letter begins: "The hype about feathered dinosaurs in the exhibit currently on display at the National Geographic Society makes the spurious claim that there is strong evidence that a wide variety of carnivorous dinosaurs had feathers. A model of the undisputed dinosaur *Deinonychus* and illustrations of baby tyrannosaurs are shown clad in feathers, all of which is simply imaginary and has no place outside of science fiction."

He continues: "The idea of feathered dinosaurs and the theropod origin of birds is being actively promulgated by a cadre of zealous scientists acting in concert with certain editors at *Nature* and *National Geographic* who themselves have become outspoken and highly biased proselytizers of the faith. Truth and careful scientific weighing of evidence have been among the first casualties in their program, which is now fast becoming one of the grander scientific hoaxes of our age."

Unfortunately, Olson's wise words fell on deaf ears. From 1999 to the present day, the National Geographic Society has never stopped peddling evolutionary science fiction as fact.

Prayer: Heavenly Father, thank You for scientists who are not afraid to speak their mind when they see science fiction being presented as fact. Amen.

Ref: http://dml.cmnh.org/1999Nov/msg00263.html.

Do Americans Still Believe in a Creator?

> *Matthew 9:38*
> *"Pray ye therefore the Lord of the harvest, that he will send forth labourers into his harvest."*

Considering the decades of evolutionary indoctrination going on in schools and the media, you might think that most Americans no longer believe in a Creator. But that's not what a recent survey from LifeWay Research reveals.

In a phone survey of 1,000 adults, 72 percent of Americans overall and 46 percent of those who don't identify with any religion agree with the survey statement: "Since the universe has organization, I think there is a creator who designed it." This view is most strongly held by evangelicals and by older adults. And most respondents – 79 percent overall, and 43 percent of those with no religious affiliation – agreed with the statement: "The fact that we exist means someone created us."

According to Mary Jo Sharp, assistant professor of apologetics at Houston Baptist University, "The infinitesimal odds that life arose by blind chance is a formidable argument" for a creator.

Ed Stetzer, executive director of LifeWay Research, said that the findings are newsworthy because the study shows that nonreligious people believe the same thing as Christians at a surprisingly high rate. Stetzer said, "This points to the possibility of a lot of conversations – if Christians would just have those conversations – telling people about the creator that they see in creation."

We couldn't agree more. Think of these Creation Moments broadcasts as conversation starters to help you tell people that there is, indeed, a Creator. And then tell them that this same Creator entered His creation and died on the cross so that their sins might be forgiven.

> *Prayer: Heavenly Father, help me learn more about creation so I can use this to introduce people to my Creator and Savior! In Jesus' Name. Amen.*

Ref: Cathy Grossman, "Religious or not, many Americans see a creator's hand," 10/8/15, Religion News Service.

Easter Island Heads –
Not Just Another Pretty Face

2 Timothy 2:15
"Study to shew thyself approved unto God, a workman that needeth not to be ashamed, rightly dividing the word of truth."

Today on Creation Moments we bring you a bit of old news. I mean *really* old news. Though this news is nearly a hundred years old, few people even today know that the massive stone heads on Easter Island are actually sitting atop tall stone bodies buried in the ground.

The statues – carved from volcanic rock between AD 1100 and 1500 and put in place by Polynesians on Easter Island – range in size from 3 feet to a towering 33 feet in height. The largest statue – named El Gigante – tips the scales at about 165 tons.

In 1919 pictures of the first excavations to Easter Island revealed that some of the statues were full sized. But over the following decades, the discoveries were gradually forgotten. Since 2010, the Easter Island Statue Project has been excavating two of the buried bodies. According to project director Jo Ann Van Tilburg, most people think the statues are only heads because there are 150 of them buried up to the shoulders on the slope of a volcano, "and these are the most famous, most beautiful and most photographed of all the Easter Island statues."

Think of it – if archaeologists had never dug deeper, they would not have uncovered the whole truth about these massive stone statues. Similarly, if we want to get a more complete understanding of what the Bible teaches, we need to dig deeper. In-depth Bible study takes time and work. But it is well worth the effort!

Prayer: Heavenly Father, I desire to know You better. Help me to make diligent study of the Bible a top priority in my life. In Jesus' Name. Amen.

Ref: Natalie Wolchover, "Do the Easter Island Heads Really Have Bodies?" *LiveScience*, 9/25/12.

"No Real Scientist Believes in Creation"

Colossians 3:23
"And whatsoever ye do, do it heartily, as to the Lord, and not unto men;"

How often have we heard the statement – "No real scientist believes in creation" – as we talk to people about why creation is actually more credible than evolution. If you ever hear someone say something like that, you'll want to listen closely to what I am about to share with you today.

When Ph.D. scientist Russell Humphreys was asked to estimate how many practicing scientists believe in biblical creation, he said: "It's a conservative estimate that there are in the U.S.A. alone around 10,000 practicing scientists who are biblical creationists." And he said this over fifteen years ago. This means there may now be many thousands more.

In the book *Search for the Truth,* author Bruce Malone lists many esteemed scientists who are biblical creationists. In addition to Dr. Humphreys, Malone mentions: Dr. Danny Faulkner, who has a Ph.D. in astronomy; Dr. John Baumgardner, with a Ph.D. in geophysics and space physics; Dr. Andrew Snelling, who has a Ph.D. in geology; and Dr. Eric Norman, a pioneer in vitamin B12 research and who holds a Ph.D. in biochemistry.

I don't have time today to tell you about other highly credentialed scientists mentioned in Malone's book who believe in creation. Each and every one has published numerous papers in journals related to their fields of expertise. And each one has said some startling things about the young age of the Earth that I will share with you on our next broadcast. You won't want to miss it.

Prayer: Heavenly Father, I pray that You will draw many more creationists into the sciences so they might glorify Your name and show evolutionists that real scientists do, in fact, believe in biblical creation! Amen.

Ref: Bruce Malone, "Creation Scientists Are Abundant," *Search for the Truth*, p. V-16 (Search for the Truth Ministries, 2001).

Esteemed Creationists Defend a Young Earth

Genesis 2:1-2
"Thus the heavens and the earth were finished, and all the host of them. And on the seventh day God ended his work which he had made; and he rested on the seventh day from all his work which he had made."

On our previous Creation Moments program, we told you about several esteemed scientists with advanced degrees who believe in biblical creation. Today I'm going to tell you what some of these scientists – and others – have said about a recent creation. First up is Dr. Russell Humphreys, who holds a Ph.D. in physics. He said that the facts of science "support a recent creation and go strongly against the idea of billions of years which the theistic evolutionists uphold."

Dr. Danny Faulkner, with a Ph.D. in astronomy, said he believes "the universe is 6,000 to 8,000 years old" and that "we have a very clear indication from scripture that creation took place in six ordinary days." Dr. John Baumgardner, with a Ph.D. in geophysics, said this: "To make sense of the world as the Bible lays it out does not allow for millions of years but does require that there be a catastrophe which destroyed all air-breathing land life except for that preserved on Noah's ark."

Finally, with a Ph.D. in chemistry, Dr. Jonathan Sarfati said, "Ninety percent of the methods that have been used to estimate the age of the earth point to an age far less than the billions of years asserted by evolutionists."

As you can see, each scientist comes from a different scientific discipline, and they all accept the Bible's account of a recent creation without hesitation. And there are thousands of other scientists in the U.S. and around the world who agree with them!

Prayer: Father, I pray that You will remind Your people that it is not shameful to believe in a recent creation. Indeed, there is no shame in believing everything that You have revealed in Your written word! Amen.

Ref: Bruce Malone, "Creation Scientists Are Abundant," *Search for the Truth*, p. V-16 (Search for the Truth Ministries, 2001). The original sources for all of these quotes can be found in Malone's book.

Creation Truth in a Comic Strip

Psalm 146:1-2
"Praise ye the LORD. Praise the LORD, O my soul. While I live will I praise the LORD: I will sing praises unto my God while I have any being."

I admit it. I enjoy reading comic strips from time to time. I especially love the single-panel "Far Side" cartoons by Gary Larson. He's one of the only cartoonists who has repeatedly focused on the foolishness of evolution and gotten away with it.

But perhaps my favorite cartoonist is the late Johnny Hart. He is famous for his award-winning "B.C." and "Wizard of Id" comic strips. Chuck Colson once referred to Hart as "the most widely read Christian of our time."

In one of my favorite comic strips, a character named Wiley sits under a tree, writing a poem called "The Seed." Let me read some of it to you.

"A seed is such a miraculous thing," Wiley writes. "It can sit on a shelf forever, but how it knows what to do, when it's stuck in the ground, is what makes it so clever. It draws nutrients from the soil through its roots and gathers its force from the sun."

Wiley then writes about how the seed reproduces itself, and how the new plant receives everything it needs from its environment. In the last panel, Wiley writes, "Perhaps all of this is a product of love." And finally he wonders: "And perhaps it happened by chance" ... which he quickly answers, "Yeah, sure."

Imagine that – creation truth in a comic strip! Hart often sparked controversy by incorporating overtly Christian themes and messages like this into his work. Perhaps he set an example that all Christians should follow!

> ***Prayer: Heavenly Father, even if I'm not in full-time ministry, I know that You want me to share biblical truth with the people I come in contact with. Lord, give me wisdom and boldness to testify what Jesus has done in my life. Amen.***

Ref: B.C. comic strip reproduced in Bruce Malone's book *Search for the Truth*, p. A-13 (Search for the Truth Ministries, 2001).

"Leaves of Three, Let Them Be"

1 John 1:7
"But if we walk in the light, as he is in the light, we have fellowship one with another, and the blood of Jesus Christ his Son cleanseth us from all sin."

"Leaves of three, let them be." "One, two, three ... don't touch me." Do you know what I'm talking about? I'm talking about poison ivy, a plant with three leaves connected closely together on each stem.

There's a good reason why poison ivy is the most well-known toxic plant in North America. Most poisons that work by touch are usually quick to lose their effect, but poison ivy leaves are covered with a sticky oil that makes the poison cling to skin and clothing. And that poison can cause a really nasty allergic reaction.

But there are even more folk sayings that are useful in identifying poison ivy. "Longer middle stem; stay away from them." This points out that the middle leaf of the three has a long stem while the two side leaves attach almost directly. "Side leaflets like mittens, will itch like the dickens." The side leaves are smaller and may be shaped like mittens. "Red leaflets in the spring, it's a dangerous thing" and "berries white, run in fright" can also help you steer clear of poison ivy.

When you think about it, sin can be compared to poison ivy. Like poison ivy, sin is fairly easy to recognize. We have the Bible's admonitions, for one. And we can count on the Holy Spirit to warn us when we are tempted to sin. Best of all, though, we have Jesus, who gives us power to resist sin and then forgives us when we fall into sin.

Prayer: *Heavenly Father, thank You for sending Your Son to die on the cross to rescue Your people from the power, the punishment and eventually the presence of sin. Amen.*

Ref: Some of this material came from "How to Identify Poison Ivy" at the wikiHow website.

A Scientific Dissent from Darwinism

Psalm 31:24
"Be of good courage, and he shall strengthen your heart, all ye that hope in the LORD."

Since Creation Moments stands firmly on the side of creation as described in the Bible, we disagree at times with the Discovery Institute, the well-known leader of the Intelligent Design movement. But we do have to recognize and applaud them for putting together a list of several hundred scientists who do not agree with Darwinism.

I'm talking about the "Scientific Dissent from Darwinism" list. To get on the list, scientists had to agree with the following statement: "We are skeptical of claims for the ability of random mutation and natural selection to account for the complexity of life. Careful examination of the evidence for Darwinian theory should be encouraged."

One such scientist – Dr. Stanley Salthe – is quoted as saying: "Darwinian evolutionary theory was my field of specialization in biology. Among other things, I wrote a textbook on the subject thirty years ago. Meanwhile, however, I have become an apostate from Darwinian theory and have described it as part of modernism's origination myth."

According to the Discovery Institute, they launched the list in 2001 because "the public has been assured that all known evidence supports Darwinism and that virtually every scientist in the world believes the theory to be true. The scientists on this list dispute the first claim and stand as living testimony in contradiction to the second."

Now ask yourself this: If these scientists were courageous in standing up for what they believe, can you – a believer in Jesus Christ – do anything less?

Prayer: Heavenly Father, I pray that Your Holy Spirit will embolden me to share the gospel with people I have been reluctant to talk to about spiritual things. In Jesus' Name. Amen.

Ref: The full list can be viewed and downloaded at http://www.dissentfromdarwin.org.

Mortimer Adler's Anti-Darwinism Crusade

Proverbs 14:6
"A scorner seeketh wisdom, and findeth it not: but knowledge is easy unto him that understandeth."

Many atheists and evolutionists claim that creationists are anti-intellectuals or know-nothing "flat-earthers." Well, let's see how well that claim holds up by looking at the life of one man.

Are you familiar with the name Mortimer Adler? No? Well, I'm quite sure you have heard of the *Encyclopedia Britannica*. For many years, Adler was chairman of the encyclopedia's board of editors. He is also the man behind the best-selling 54-volume *Great Books of the Western World*.

The reason I'm telling you about Adler today, however, has to do with his half-century-long crusade against Darwinism. Based on his understanding of what constitutes real science, Adler contended that Darwinism was nothing but a myth and wild speculation. Adler observed, "Darwin himself is partly responsible for much of this speculation. *The Origin of Species* is full of guesses which are clearly unsupported by the evidence."

Adler added that "these guesses ... are not in the field of scientific knowledge anyway.... Evolution is not a scientific fact, but at best a probable history, a history for which the evidence is insufficient and conflicting."

In his book, *Slaughter of the Dissidents*, Dr. Jerry Bergman points out that Adler was born of Jewish parents and became a "religious scoffer" when he was a young man. However, it was his acceptance of the cosmological argument for God – that a creation needed a Creator – that eventually caused Adler to believe in God.

Mortimer Adler died in 2001 at the age of 98 – a Darwin-denier until the end.

Prayer: Heavenly Father, thank You for opening the eyes of millions to the fact that Darwinism is not a fact at all! In Jesus' Name. Amen.

Ref: Dr. Jerry Bergman, *Slaughter of the Dissidents, Volume 1*, pp. 277-281 (Leafcutter Press, Second Edition, 2012).

Darwinism Rests Its Case on a Lawyer's Claims

Luke 19:40
"And he answered and said unto them, I tell you that, if these should hold their peace, the stones would immediately cry out."

In the introduction to his chapter on the origin of life for the book *Evolution's Achilles' Heels,* Ph.D. scientist Tas Walker points out that "geology has long been considered the solid foundation for evolution." He goes on to write that Darwin "built his theory of biological evolution squarely on then-current theories of geology, especially the work of his contemporary, Charles Lyell."

Now, we've told you about Charles Lyell on previous broadcasts. But today it's important to stress that Lyell was trained to be a lawyer, not a scientist. So, Dr. Walker notes, "Often without knowing it, geologists owe their current old-age position to the lawyer turned uniformitarian geologist, Lyell."

The second thing Dr. Walker points out is that Lyell was a man with an agenda. Walker writes, "Lyell was clear about what he wanted to achieve, as he revealed to a colleague in his correspondence. He said his aim was 'to free the science [of geology] from Moses.'"

The rest of the chapter is filled with geological evidences supporting a young Earth. But he began his chapter talking about Lyell because "the so-called geologic record has given the historical sciences an earth 'history' of billions of years that allows unobservable biological evolution to seem plausible. Without eons of time," Dr. Walker concludes, "evolution would be dead in the water." Not only is it dead in the water. When you examine the geological evidence, you see that the stones really do cry out they were created by Jesus Christ.

Prayer: Heavenly Father, I know that Darwinism is built on a foundation of assumptions fueled by a hatred of You. I pray that You will open the eyes of many who are now blindly following men like Lyell and Darwin. Amen.

Ref: Tasman Walker, Ph.D., "The Geological Record," *Evolution's Achilles' Heels,* pp. 155-160 (Creation Book Publishers, 2014).

Could a Magnet Separate You from God?

John 6:39
"And this is the Father's will which hath sent me, that of all which he hath given me I should lose nothing, but should raise it up again at the last day."

According to the *ScienceDaily* website, "New research involving a psychologist from the University of York has revealed for the first time that both belief in God and prejudice towards immigrants can be reduced by directing magnetic energy into the brain."

The article goes on to say that the psychologist led a team of researchers from UCLA to "carry out an innovative experiment using transcranial magnetic stimulation, a safe way of temporarily shutting down specific regions of the brain." The findings revealed that "people in whom the targeted brain region was temporarily shut down reported 32.8% less belief in God, angels, or heaven." They also discovered they could use the same technique to turn off a person's patriotism.

The head of the research team noted: "As expected, we found that when we experimentally turned down the posterior medial frontal cortex, people were less inclined to reach for comforting religious ideas despite having been reminded of death."

Dr. Colin Holbrook from UCLA noted that the findings "are very striking" and "consistent with the idea that brain mechanisms that evolved for relatively basic threat-response functions are repurposed to also produce ideological reactions."

Well, I wouldn't worry about strong magnets taking away the faith of those who have put their trust in Jesus. As it says in the Bible, our Savior will not lose even one of His precious saints but will deliver them all to His Father in heaven.

Prayer: Heavenly Father, thank You for not only saving me but for keeping me until the end! In Jesus' Name. Amen.

Ref: University of York in the UK, "Belief in God and prejudice reduced by directing magnetic energy into the brain," *ScienceDaily*, 10/14/15.

"No, Thanks. I'd Rather Walk."

Deuteronomy 13:4
"Ye shall walk after the LORD your God, and fear him, and keep his commandments, and obey his voice, and ye shall serve him, and cleave unto him."

As everyone knows, most sea creatures get from place to place by swimming. Others – like the octopus – use a form of jet propulsion. But one thing that no one expects to see on the sea floor is a fish that walks! And what makes this creature even more unusual is that the spotted handfish gets about on pectoral fins that look almost like human hands!

This unusual form of locomotion isn't the only almost-human thing about the spotted handfish. According to an article in *ScienceNews*, they "fish the way people do, with patience and lures." In their words, "The fishes' dorsal fins' front spines have evolved flamboyant deceptions: tendrils that wiggle like worms, a pom-pom on a stick, a lump with stripes and an eyelike spot."

Once these lures bring the fish's lunch within range, in a flash the fish opens its mouth, creating a suction that sucks the prey into the fish's mouth.

Despite its skills at fishing, the spotted handfish has become one of the most endangered creatures on Earth. Efforts are underway to keep them from total extinction. Let me add that our Creator has also made considerable efforts of His own for rescuing another kind of walking creature from a fate even worse than extinction. Humans have disobeyed God and refused to walk in His ways. Thankfully, our Creator entered His creation and died on the cross to rescue us from a fate far worse than death. Oh, what a Savior!

> ***Prayer: Heavenly Father, the Bible tells us that we are like sheep that have gone astray. We have turned every one to his own way ... but You placed on Jesus the iniquity of us all. Thank You for such great salvation! Amen.***

Ref: Susan Milius, "These fish would rather walk," *ScienceNews*, 10/3/15, pp. 4-5.

Coffee – Brimming with Health Benefits

Genesis 1:12
"And the earth brought forth grass, and herb yielding seed after his kind, and the tree yielding fruit, whose seed was in itself, after his kind: and God saw that it was good."

    I'm not much of a coffee drinker, but after hearing about its many health benefits, I just might make coffee my drink of choice.
    According to a cover story in *ScienceNews*, "Scientific findings in support of coffee's nutritional attributes have been arriving at a steady drip since the 1980s, when Norwegian researchers reported that coffee seemed to fend off liver disease." Since then, coffee has shown value against liver cancer, type-2 diabetes, heart disease and stroke. It also appears to protect against depression, Parkinson's and Alzheimer's diseases. In fact, I don't even have time on today's program to list all of coffee's health benefits.
    The latest studies show that people who drink two or more cups of coffee a day live longer than those who don't. In a recent Japanese study of more than 90,000 people, those who drank three to four cups a day were 24 percent less likely to die during the next nineteen years than people who didn't drink coffee at all.
    Think about it – God created the coffee bean on day three of Creation Week, before Adam and Eve sinned and brought disease and death into the world. Is it possible that God, knowing that Adam would sin, provided a remedy for many of the diseases that would come about because of that sin? I believe He did. He also knew long before He created the world that you and I would need a Savior. Is Jesus Christ *your* Savior? If not, put your trust in Him today!

Prayer: Heavenly Father, You have provided everything we need for this life and the next. Thank You especially for providing Your Son to die on the cross in our place to make us blameless in Your sight. In Jesus' Name. Amen.

Ref: Nathan Seppa, "The Beneficial Bean," *ScienceNews*, 10/3/15, pp. 16-19.

The Evolution Illusion

1 Corinthians 1:20
"Where is the wise? where is the scribe? where is the disputer of this world? hath not God made foolish the wisdom of this world?"

Evolutionary biologists like to give people the impression that they are unbiased individuals who follow the evidence wherever it leads. If only this were true!

As pointed out in the book *In the Minds of Men,* ever since the publication of Darwin's *Origin of Species* in 1859, scientists have consistently interpreted natural phenomena in a way that appears to provide evidence to support the theory of evolution. Some of these interpretations have turned out to be based on faulty observation, some on faulty reasoning, and some on blatant fraud. When you look at the so-called evidence up close, you'll see that evolution is just an illusion.

Darwin's theory of evolution has, in many minds, displaced the biblical creation account of our origins. For those who hold to evolution, it is vitally important to cling to any evidence for evolution, no matter how flawed that evidence might be. For the same reason, there is an extreme reluctance in the scientific community to even consider new evidence that does not support evolution.

The idea that scientists think rationally and fairly is a simplistic myth. Looking inside the ivory towers of science, we find the familiar power establishments, personality conflicts and intellectual blind spots that are present in any other profession. In other words, they are fallen human beings who are unwilling to be illuminated by the light that emanates from every page of God's Word.

Prayer: Heavenly Father, let me never put scientists on a pedestal or elevate them to be Your equal. Man's wisdom will fail. In You alone is found knowledge, wisdom and truth. In Jesus' Name. Amen.

Ref: Ian Taylor, *In the Minds of Men: Darwin and the New World Order*, pp. 278-280 (TFE Publishing/Creation Moments, Sixth Edition, 2008). Book is available for purchase from Creation Moments.

Virtual Reality – Blessing or Curse?

Ecclesiastes 2:1
"I said in mine heart, Go to now, I will prove thee with mirth, therefore enjoy pleasure: and, behold, this also is vanity."

While visiting Chicago's Museum of Science and Industry about twenty years ago, a Creation Moments staff member stood in a long line, waiting to take his first look at what we now know as "virtual reality." Today you can experience virtual reality – or VR for short – on your smartphone with a headset that can cost as little as ten dollars.

When you strap the glasses onto your head, your eyes are presented with two slightly different images, which your brain fuses into a single three-dimensional image. This is far beyond anything you experienced as a child, looking at 3D photos with that plastic toy that was so popular back then. While wearing VR glasses, you can look up and down and all around you. You can skydive, go deep-sea diving and walk on the moon, all without leaving your home.

As exciting as virtual reality may sound, it is not without its risks. As futurist Ray Kurzweil predicts, "By the 2030s, virtual reality will be totally realistic and compelling, and we will spend most of our time in virtual environments.... We will all become virtual humans."

This is truly escapism brought to a new level. With all of life's pain and problems, we can certainly understand why Kurzweil and others are eager to escape from reality. However, virtual reality – just like alcohol, drugs and over-indulging in sports or entertainment – ultimately fails to provide relief that lasts. Only Jesus can fulfill our deepest needs, both now and in eternity.

> *Prayer: Heavenly Father, You have given mankind the intelligence to invent exciting, new technologies. Though I am free to enjoy these things, remind me to enjoy them in moderation. In Jesus' Name. Amen.*

Ref: Monica Kim, "The Good and the Bad of Escaping to Virtual Reality," *The Atlantic*, 2/18/15.

Niagara Falls – Evidence for a Young Earth

John 4:14
"But whosoever drinketh of the water that I shall give him shall never thirst; but the water that I shall give him shall be in him a well of water springing up into everlasting life."

While visiting Niagara Falls in 1840, Charles Lyell was told that the falls were receding at about three feet per year. This was based on what eyewitnesses who lived in that region had reported. But for reasons Lyell did not disclose, this lawyer-turned-geologist concluded that a "much more likely *conjecture*" – his words, not mine – is that the rate of erosion was only one foot per year. At this rate, it showed that Niagara Falls had been flowing for 35,000 years. Now, why would Lyell make such a conjecture?

Dr. Henry Morris answers that question in his book *The Long War Against God*. "Lyell's work at Niagara," wrote Morris, "accomplished its main goal – that of calling Scripture into question. For biblical chronology cannot allow 35,000 years since Noah's Flood. And if Genesis is wrong, how can we trust any other portion?"

So how fast is Niagara Falls *really* receding? Though there is some dispute over this, I believe that the most accurate measurements show the rate of recession to be over six feet per year. This means that the age of Niagara Falls turns out to be less than 6,000 years, a near confirmation of the date proposed by Bishop Ussher for the age of the Earth.

Since Darwin's theory of evolution requires a very ancient Earth – and he used Lyell's writings to support that view – where does that leave evolution? Truly, the more one knows about evolution and its anti-God agenda, the less scientific it appears.

Prayer: Heavenly Father, thank You for evidences that support the truth that Your Word is accurate in everything it describes. In Jesus' Name. Amen.

Ref: Henry Morris, *The Long War Against God,* p. 189 (Master Books, 1989). See also Ian Taylor, "The Age of the Earth," http://www.creationmoments.com/content/age-earth.

Eau de Whale Guts

2 Corinthians 2:15
"For we are unto God a sweet savour of Christ, in them that are saved, and in them that perish:"

As one perfume expert puts it, "It's beyond comprehension how beautiful it is. It's transformative. Its like an olfactory gemstone."

The perfumer was describing ambergris, a waxy excretion found in the intestines of sperm whales. Since most countries have made it illegal to slaughter sperm whales, ambergris is very hard to come by. In fact, it is almost worth its weight in gold. As one beachcomber observed, "There aren't too many professions where you could go to work and stumble upon $30,000 one morning."

Bloomberg Business writes, "To outsiders, it may seem like easy money – ambergris can wash ashore anywhere there are sperm whales." Ironically, many ambergris hunters don't even know which whale orifice it comes from. As neuroscientist Chris Kemp said, "Despite what most people think, it is not vomit. That's one of the biggest misnomers about ambergris. Unfortunately, it comes out the other end."

Sure enough, ambergris smells like perfumed cow dung. What makes ambergris so special is that it *amplifies* the other fragrances found in perfumes. As one perfumer noted, "It alters the quality of the existing notes and makes them bigger, deeper and more expansive than they can ever be on their own."

Unlike fragrant perfumes, sinful man has nothing within himself that smells good before a holy God. And if our natural scent is amplified, we only smell worse. But when Christ enters our lives, we begin to give off the sweet-smelling aroma of Christ to others. Best of all, we smell good to God, too!

Prayer: Heavenly Father, thank You for sending Jesus to replace the foul stench of a sinner with the wonderful aroma of a new creature in Christ! In Jesus' Name. Amen.

Ref: Eric Spitznagel, "Ambergris, Treasure of the Deep," Bloomberg Business, 1/12/12.

Creation Adventures Near You

Proverbs 8:32-33
"Now therefore hearken unto me, O ye children: for blessed are they that keep my ways. Hear instruction, and be wise, and refuse it not."

Are you hungry to learn more about God's magnificent creation and teach your family what they need to know to defend the faith from the pseudoscience of evolution? Of course, the Creation Moments website is a great place to start.

But there are other ways to learn about biblical creation – including outstanding books and DVDs on the six days of creation, the worldwide flood and so much more. Another great way is to visit the big Creation Museum near Cincinnati. But what if a trip there can't be worked into your family's vacation plans? You can visit one of the many other creation museums scattered all across the United States and in other countries around the globe.

So how do you find out where these museums are located and what each one has to offer? The Creation Science Alliance – of which Creation Moments is a member – is a network of creation ministries and museums around the world, and they have put together the "Big Map of Creation" that makes it easy to find creation tours and museums near you. Simply point your web browser to VisitCreation.org to get started.

On this map, you are sure to find sound doctrine and dependable science at any location your family chooses to visit, and you'll see fossils and dinosaur bones up close. These smaller museums are also great because your family will receive personal attention from trained creation tour guides.

We highly recommend a creation museum tour for your next vacation.

Prayer: Heavenly Father, thank You for blessing Your people with so many opportunities to learn about Your wonderful creation! In Jesus' Name. Amen.

Ref: www.VisitCreation.org.

Do Lemmings Commit Suicide?

Genesis 9:1
"And God blessed Noah and his sons, and said unto them, Be fruitful, and multiply, and replenish the earth."

True or false? When the population of lemmings grows too large, they get into a frenzied state, and huge groups of the furry rodents commit suicide by hurling themselves into the nearest body of water.

Many people think this is true, but it is really nothing but a, well, *rural* legend that might have gotten its start from a news report that came out of Norway in 1868. A steamer there reportedly sailed for 15 minutes through a swarm of swimming and drowning lemmings. According to the report, the swarm was two-to-three miles wide.

Actually, lemmings really do throw themselves into the water – but not to commit suicide. They are simply migrating after depleting food supplies in their area. During a mass migration, many lemmings do, in fact, die from drowning. But more lemmings are eaten by predators.

Today there is another overpopulation legend that far too many people are convinced is true. It's the notion that mankind is overpopulating the planet. Some environmentalists have called for reduction of world population to just two billion – a reduction of over two-thirds of the world population. Others would like to see only a half billion people on our planet. And many think it's permissible to kill humans in large numbers to avoid running out of food and other vital resources.

No, overpopulation is not the problem. The real problem is that man, in his pride, thinks he knows more than God knows about what is best for the planet that He created.

Prayer: Heavenly Father, many environmentalists today stand in opposition to Your clear command to be fruitful and multiply. I pray that You will show them that Your ways are always the best! In Jesus' Name. Amen.

Ref: "Leaping lemmings," *Secrets of the Natural World*, p.70 (Readers Digest/Dorling Kindersley, 1993).

Did Our Hands Evolve for Fighting?

Genesis 4:8
"And Cain talked with Abel his brother: and it came to pass, when they were in the field, that Cain rose up against Abel his brother, and slew him."

Today on Creation Moments we bring you a rather bizarre story. Its seems that scientists used the arms and hands of corpses in an attempt to show that human hands evolved for fighting. True story. In fact, it was reported in the *Journal of Experimental Biology*.

The scientists did this to see if a clenched fist is more effective at protecting the metacarpal bones of human hands when we punch something ... or someone. Not surprisingly, they found that humans can safely strike with 55 percent more force with a clenched fist rather than slapping someone with the thumb extended and loosely folded fingers.

This is pretty obvious when you think about it. What's not so obvious is how the researchers concluded that evolution had anything to do with it. All they could really conclude is that even the earliest of men had a violent streak.

This should not surprise us at all because the Bible tells us that Adam and Eve's son Cain struck his brother and killed him ... and there is no record of him breaking his metacarpals. Obviously, we have always had the innate tendency to clench our fists for fighting. And as history has taught us, we've done that way too often.

While the fall of Adam is responsible for man's violent streak, Jesus Christ teaches us to love all men, even our enemies. Rather than forming our hands into fists, let us use them to do good unto believers and unbelievers alike.

Prayer: Heavenly Father, whenever I feel anger rising up inside me, remind me that You want me to love my enemies. In Jesus' strength, I know I can do that! Amen.

Ref: "Dead Man Punching Sheds Light on Fist Evolution," *Discovery News*, 10/21/15.

Did Our Faces Evolve for Being Punched?

Genesis 6:11
"The earth also was corrupt before God, and the earth was filled with violence."

On our previous Creation Moments broadcast, we told you about researchers who are claiming that our hands evolved so that they can be formed into fists for fighting. Now that same group of researchers is also telling us that our faces evolved to withstand being punched.

This idea represents a big departure from the long-held theory that human faces look the way they do because our ancestors spent much of their time chewing nuts and other hard foods. The lead author of the study writes, "The australopiths were characterized by a suite of traits that may have improved fighting ability, including hand proportions that allow formation of a fist." He adds, "If indeed the evolution of our hand proportions were associated with selection for fighting behavior, you might expect the primary target, the face, to have undergone evolution to better protect it from injury when punched."

A member of the research team added: "Our research is about peace. We seek to explore, understand, and confront humankind's violent and aggressive tendencies. Through our research we hope to look ourselves in the mirror and begin the difficult work of changing ourselves for the better."

Good luck with that! If history has taught us anything, it's that mankind – starting with Cain – has had a violent streak and, if anything, we are getting worse. Only God can take such violence out of our hearts when we turn to Christ for the forgiveness of sins.

Prayer: Heavenly Father, I pray that Your Holy Spirit will help me control my temper. May my actions always be a silent testimony to the new creature in Christ You have made me to be! In Jesus' Name. Amen.

Ref: Jennifer Viegas, "Human Face Evolved to Withstand Punching," *Discovery News*, 6/9/14.

Metal That's Light as a Feather

Matthew 11:29-30
"Take my yoke upon you, and learn of me; for I am meek and lowly in heart: and ye shall find rest unto your souls. For my yoke is easy, and my burden is light."

HRL Laboratories, owned by Boeing, startled the world years ago when they unveiled the lightest metal ever produced. More recently they're in the news again after the parent company released a video talking about what *Popular Mechanics* called one of the top 10 world-changing innovations of 2012.

I'm referring to microlattice – a metal that is so light, fluffy white dandelion seeds can hold it up. According to *Popular Mechanics*, the metal is a hundred times lighter than Styrofoam packing peanuts and could be useful for medical applications and in the automotive and aerospace industries.

What makes this metal so light is answered in the Boeing video. HRL Laboratories' Sophia Yang compares microlattice to the basic composition of bones. While the outer structure of our bones is rigid and solid, the interior is filled with thin, lattice-like tissue. The outer walls of microlattice are only about one-one thousandth the width of a human hair.

As one news website correctly points out, "Bones are both strong and light due to this composition, and microlattice works on the same basic principle. It is made up of 99.99 percent air, so it's extremely light, but it has remarkable compression properties."

We have posted a link to Boeing's video at the Creation Moments website. It's a good video, but we would have liked it even more if Boeing had given credit to our Creator for designing the bones that inspired HRL Labs to design the world's "lightest" metal!

Prayer: Heavenly Father, though evolutionists say that the designs we see in nature only appear to have been designed, I pray that more and more people will come to see how foolish such a statement really is. In Jesus' Name. Amen.

Ref: Jeremy Laukkonen, "Lightest Metal Ever, Boeing's Microlattice Weighs Less Than Styrofoam [Video]," Inquisitr, 10/13/15. Video: http://www.inquisitr.com/2493098/lightest-metal-ever-boeings-microlattice-weighs-less-than-styrofoam-video/

What Your Tears Reveal About You

Revelation 21:4
"And God shall wipe away all tears from their eyes; and there shall be no more death, neither sorrow, nor crying, neither shall there be any more pain: for the former things are passed away."

Did you know that your tears reveal a whole lot more about you than what you might expect?

Photographer Rose-Lynn Fisher discovered what our tears reveal about us when taking pictures of the dried tears of one hundred subjects for her project called "The Topography of Tears." When viewing her own tears, Fisher said they "looked like an aerial view, almost as if I was looking down at a landscape from a plane."

The Telegraph news website points out that "tears contain oils, antibodies and enzymes and fall into three categories; basal, which are released continuously to keep the eyes lubricated; reflex, which occur in response to irritants such as when chopping onions or when getting poked in the eye; and psychic, triggered by emotions."

The website also noted that "scientists have identified that different types of tears are made up of distinct molecules. For example, those caused by emotions contain hormones which act as a painkiller and are released when we are stressed."

Wasn't that good of our Creator to supply us with tears to help us deal with both emotional and physical pain? Even better, He is always ready to comfort us in times of sorrow. Tears are not the result of mutations over millions of years. They were given to us by a Creator who experienced great emotional pain in the garden of Gethsemane and incredible physical pain when He died on the cross for you and me.

Prayer: Heavenly Father, thank You for giving us tears that can do so much more than keep our eyes clean and moist. I am grateful that You will wipe every tear from our eyes when we are in Your presence. Amen.

Ref: Claire Carter, "Structure of tears influenced by what makes us cry," *The Telegraph*, 5/16/14.

Those Strange Nazca Lines

Genesis 4:22a
"And Zillah, she also bare Tubalcain, an instructer of every artificer in brass and iron:"

I'm sure that many of you have seen aerial photos of massive drawings that can only be viewed while flying over the Nazca Desert south of Lima, Peru. Some of these images depict spiders, hummingbirds and monkeys. Others are perfectly straight lines that run as far as five miles. In fact, writes Dr. Donald Chittick in *The Puzzle of Ancient Man*, "the lines are as straight as our best modern methods of aerial surveying could make them."

Scholars believe the lines were created by the Nazca culture between 500 BC and 500 AD. And this is why most archaeologists find these figures so perplexing. They assume that the Nazca people were too primitive to have known anything about surveying techniques. But as Dr. Chittick points out, "Evidence indicates early surveying was available to the builders of the pyramids in Egypt," and the Nazca people "probably knew about surveying methods" as well.

The question remains of why the Nazca people created massive figures in the desert that could only be seen from an airplane or hot-air balloon. Dr. Chittick points out that researchers have discovered that the Nazca people knew how to make a high-quality black cloth with a very fine weave that could have been used for constructing hot-air balloons. The black cloth would absorb the rays of the sun and heat the air inside.

Sound far-fetched? Considering that the Bible tells us that even the earliest men were building cities and making things of brass and iron, it isn't implausible at all.

Prayer: Heavenly Father, thank You for the Bible, which gives us a better understanding of the world around us and reveals how sinful man can be adopted into Your family through saving faith in Christ Jesus. Amen.

Ref: Donald E. Chittick, Ph.D., *The Puzzle of Ancient Man*, pp. 177-188 (Creation Compass, Third Edition, 2006).

The Spider That Thinks It's a Scuba Diver

Genesis 6:20
"Of fowls after their kind, and of cattle after their kind, of every creeping thing of the earth after his kind, two of every sort shall come unto thee, to keep them alive."

Most spiders use their silk to make webs to catch prey. But there's one spider that spins its silk to make a virtually waterproof sack and fills it with air bubbles so it can survive underwater for hours at a time. In fact, the air-filled sack is so efficient, it allows this air-breathing spider to live virtually its whole life a few inches below the water's surface.

Scientists have known about the diving bell spider for a long time. But a few years ago they discovered that this silken sack is actually able to produce its own air supply. Acting much like a fish's gills, the finely woven sack draws oxygen from the water while letting nitrogen exit the diving bell.

Evolutionists can't explain how the spider learned how to scuba dive and produce silk that makes the diving bell virtually waterproof. Oh, they will say what they always say – natural selection and mutations over millions of years. But creationists have no need of such far-fetched explanations.

The Creator showed His fondness for this spider by giving it what it needs to avoid predators above the water while feeding upon underwater creatures. Now, think about how much more our Creator loves each one of us. He gave us the gift of His Son, who lived a sinless life and then died as a spotless sacrifice on our behalf so we would be found holy and blameless in God's sight through faith in Christ.

Prayer: Heavenly Father, thank You for providing such a great gift that cost You so much! I don't deserve such a gift, but I gratefully accept the love You poured out on me through Jesus Christ. Amen.

Ref: "Diving Bell Spider Uses Bubble Like Gills," Discovery News, 6/9/11.

Scientists Warn of Impending Crisis

Psalm 106:38
"And shed innocent blood, even the blood of their sons and of their daughters, whom they sacrificed unto the idols of Canaan: and the land was polluted with blood."

Best-selling author of *Jurassic Park* and many other novels, the late Michael Crichton, added a fascinating non-fiction appendix at the end of his novel *State of Fear*. Listen to how it begins:

"Imagine that there is a new scientific theory that warns of an impending crisis, and points to a way out. This theory quickly draws support from leading scientists, politicians and celebrities around the world. Research is funded by distinguished philanthropists, and carried out at prestigious universities. The crisis is reported frequently in the media. The science is taught in college and high school classrooms."

Crichton then surprises his readers: "I don't mean global warming," he writes. "I'm talking about another theory, which rose to prominence a century ago." What is this theory that – like global warming – was supported by the scientific community? He was talking about eugenics – a diabolical theory that crawled out of the twisted mind of Francis Galton, half-cousin of Charles Darwin. With its core belief that some people groups are inferior and don't deserve to live, eugenics led to the concentration camps of Nazi Germany.

A great proponent of eugenics was Margaret Sanger, founder of Planned Parenthood. She said, "Fostering the good-for-nothing at the expense of the good is an extreme cruelty. There is no greater curse to posterity than that of bequeathing them an increasing population of imbeciles." Who are these imbeciles she was talking about? Find out on our next broadcast as we look at famous people who believed this so-called scientific theory.

Prayer: Father, bless the biblical creation ministries that are informing people that Darwin's writings led to the death of countless millions through eugenics. Amen.

Ref: "Why Politicized Science Is Dangerous," Appendix 1 of Michael Crichton's novel *State of Fear*, pp. 575-576 (Harper, 2009).

Innocent Blood

Proverbs 6:16-17
"These six things doth the LORD hate: yea, seven are an abomination unto him: A proud look, a lying tongue, and hands that shed innocent blood..."

I'm sure you've heard evolutionists say that all real scientists believe in evolution. Well, let me tell you about a different theory that many if not most scientists believed in the late nineteenth and early twentieth centuries – eugenics. This is the belief that some people groups – like blacks and Jews – are inferior and should be eliminated to improve the gene pool of the human race. In theory, it made sense to many. But when the innocent blood of millions was spilled by Adolf Hitler, suddenly it didn't seem like such a good idea.

Best-selling novelist, the late Michael Crichton, points out in his book *State of Fear* that supporters of eugenics didn't just include scientists. It included statesmen like Theodore Roosevelt, Woodrow Wilson and Winston Churchill, inventor Alexander Graham Bell, author H.G. Wells, playwright George Bernard Shaw and many, many others who were among the elites of society.

Eugenics even had the support of the National Academy of Sciences and the American Medical Association. And let's not forget Margaret Sanger, founder of Planned Parenthood, an organization that continues killing millions of babies every year.

Well, as deadly as eugenics turned out to be, it doesn't even come close to the millions of lives lost because of Darwin's theory of evolution. Not only has Darwinism led to the slaughter of countless millions by ruthless dictators, it even destroys the perpetrators of such cruel acts by giving them an excuse to deny the Bible's account of creation and reject salvation in Christ.

Prayer: Heavenly Father, I pray that You will help me and biblical-creation ministries like Creation Moments expose evolution as the pseudoscience it really is. In Jesus' Name. Amen.

Ref: "Why Politicized Science Is Dangerous," Appendix 1 of Michael Crichton's novel *State of Fear*, pp. 575-577 (Harper, 2009).

"Primitive" Electric Eel Not So Primitive After All

Revelation 4:11
"Thou art worthy, O Lord, to receive glory and honour and power: for thou hast created all things, and for thy pleasure they are and were created."

Not long ago, we told you about the ingenious way that electric eels release strong electric current into the water to make fish thrash about and come out of hiding. That's a pretty cool trick from such a supposedly primitive creature.

Now scientists have discovered another shockingly sophisticated trick used by electric eels. As reported by Reuters, "Vanderbilt University neurobiologist Kenneth Catania said some have viewed electric eels as unsophisticated, primitive creatures with a single tool in the toolbox, shocking their prey to death." But he added, "In reality, they manipulate their electric fields in complex ways that only now are being appreciated."

Basically, electric eels bring their positive end – located in their head – close to the negative end in their tail. By trapping their prey between the positive and negative ends of their "battery", the eels are able to more than double the voltage inflicted on prey.

According to Catania, "We know from basic physics that bringing two electrical poles together concentrates the electric field, and we know from basic muscle physiology that running a muscle too fast for too long causes exhaustion. But I would never have imagined an electric eel could produce the same results."

Dr. Catania, do you know why you would never have imagined that? Because you believe – despite the evidence – that electric eels are primitive creatures produced by non-intelligent chance mutations. But creationists who study electric eels aren't shocked by their sophisticated abilities. We know they were created by the same God who created electricity!

Prayer: Heavenly Father, thank You that so-called "primitive" animals give powerful testimony that they are not primitive at all! In Jesus' Name. Amen.

Ref: Will Dunham, "Zap happy: electric eels innovative in subduing hapless prey," Reuters, 10/28/15.

Blood – Another Masterpiece of Design
1 Peter 1:2
"Elect according to the foreknowledge of God the Father, through sanctification of the Spirit, unto obedience and sprinkling of the blood of Jesus Christ: Grace unto you, and peace, be multiplied."

Have you ever wondered why blood cells look like red discs that are indented toward the middle? Furthermore, is this distinctive shape an element of design or is it the product of millions of years of mutations and natural selection?

Creationist and biologist Alan Gillen makes a strong case for design in his aptly named book *Body by Design*. According to Dr. Gillen, scientists used IBM mainframe computers to determine the ideal shape for a cell whose job is transporting oxygen to cells and taking away waste products. They discovered that the red blood cell's bio-concave shape allows for maximum surface contact of oxygen-carrying hemoglobin with the cells. This facilitates the two-way exchange of blood gases.

He also points out that this shape gives red blood cells flexibility and elasticity, so they can be folded when flowing through narrow blood capillaries. Plus, they have smooth, rounded edges that reduce the amount of friction as they scurry through veins, arteries and capillaries.

Now, what if evolution's first attempt produced red blood cells that were shaped like cubes and had thin, sharp edges? What if the cells didn't have their distinctive bio-concave shape? Obviously, blood cells had to be designed just right the first time or else the living organism would have died.

And while we're on the subject, try asking an evolutionist which evolved first – the blood, the heart or the circulatory system with its intricate network of veins, arteries and capillaries? Once again, God had to have created them all at the exact same time.

Prayer: Heavenly Father, while blood is necessary for life, it is the blood of Jesus poured out for our sins that makes it possible for us to have eternal life. In Jesus' Name. Amen.

Ref: Alan L. Gillen, *Body by Design*, p. 76 (Master Books, Seventh Printing, 2011).

The Fish with Three Lines of Defense

Ephesians 6:11
"Put on the whole armour of God, that ye may be able to stand against the wiles of the devil."

To hear evolutionists tell it, millions of years of chance mutations, combined with natural selection, produced all manner of living creatures. Now, let's look at just one creature – the porcupinefish – and see how the theory of evolution works out in real life. If the theory fails the test, it's a strong indicator that the theory lacks the power to explain the existence of *any* living creature.

The porcupinefish is blessed with not one, not two, but three lines of defense. The first comes into play when a predator is near. The porcupinefish quickly swallows a large amount of water, practically doubling its size. This makes the fish too large for some predators to swallow.

The second line of defense comes into play as the fish transforms itself into a football-sized balloon covered with menacing barbs. Its scales stand on end, turning it into an aquatic version of the land-dwelling porcupine. The very sight of these sharp scales protruding in every direction scares off many predators.

The third line of defense for some porcupinefish species is the neurotoxin in their internal organs – a toxin that's at least twelve-hundred times more potent than cyanide. Wikipedia correctly points out, "As a result of these three defenses, porcupinefish have few predators."

Now, can evolutionists explain how this unique fish came to have these three lines of defense? No, they can't. Not even Darwin tried to explain it when he wrote about the porcupinefish in his book *The Voyage of the Beagle*.

Prayer: Heavenly Father, thank You for not leaving me defenseless but for giving me everything I need to stand against the wiles of the devil. In Jesus' Name. Amen.

Ref: *1000 Wonders of Nature*, p. 120 (Reader's Digest, London). Wikipedia entry on "Porcupinefish."

The Implausibility of Evolution

Deuteronomy 31:6
"Be strong and of a good courage, fear not, nor be afraid of them: for the LORD thy God, he it is that doth go with thee; he will not fail thee, nor forsake thee."

If you're a long-time listener, you know that Creation Moments does not discuss politics. That's why today's program focuses not on Ben Carson's political involvements but on Dr. Carson's courage as a scientist and world-renowned surgeon for sharing his politically incorrect views on Darwinian evolution.

In his book, *Take the Risk*, Dr. Carson makes it clear that he rejects Darwinian evolution. Here is a portion of what he wrote:

"For me, the plausibility of evolution is ... strained by Darwin's assertion that within fifty to one hundred years of his time, scientists would become geologically sophisticated enough to find the fossil remains of the entire evolutionary tree in an unequivocal step-by-step progression of life from amoeba to man – including all of the intermediate species."

Dr. Carson continues: "Of course, that was 150 years ago, and there is still no such evidence. It's just not there. But when you bring that up to the proponents of Darwinism, the best explanation they can come up with is 'Well ... uh ... it's lost!' Here again I find it requires too much faith for me to believe that explanation, given all the fossils we have found without any fossilized evidence of the direct, step-by-step evolutionary progression from simple to complex organisms or from one species to another species. Shrugging and saying, 'Well, it was mysteriously lost, and we'll probably never find it,' doesn't seem like a particularly satisfying, objective, or scientific response."

Doctor Carson, we couldn't agree more!

Prayer: *Heavenly Father, thank You for scientists like Dr. Carson who take the risk of opposing Darwinism. In Jesus' Name. Amen.*

Ref: Dr. Ben Carson, *Take the Risk: Learning to Identify, Choose, and Live with Acceptable Risk,* pp. 160-161 (Zondervan, 2008).

Young Blood

Job 41:24
"His heart is as firm as a stone; yea, as hard as a piece of the nether millstone."

As dinosaur bones containing blood vessels and blood cells are continually being found, more and more laboratories are assigning dates to these bones in the thousands, not millions, of years. This makes evolutionists just like the patient who goes to his doctor, convinced that he is dead. After failing to convince the patient he is alive, the doctor pricks the man's finger, causing it to bleed. "See?" the doctor says. You must be alive because dead men don't bleed." And the patient replies: "Imagine that! Dead men *do* bleed after all!"

Evolutionists know full well that dinosaur bones that are millions of years old should not have blood and blood vessels in them. To defend their position, they usually say the bones were contaminated, throwing off the dating. But North Carolina State University said that their researchers "have confirmed that blood vessel-like structures found in an 80 million-year-old hadrosaur fossil are original to the animal, and not biofilm *or other contaminants.*"

But like the patient who kept thinking he was dead, the university went on to say that their findings "add to the growing body of evidence that structures like blood vessels and cells *can* persist over millions of years."

Evolutionists just won't admit that their evolutionary dating of dinosaurs has finally been falsified. After all, that would be like admitting that the Bible is right, and this is something that atheistic evolutionists will never do.

Prayer: Heavenly Father, without You opening the eyes of evolutionists, I know it is impossible for me to get them to believe. Open their eyes to the truths found in Your infallible Word, I pray. Amen.

Ref: "Researchers Confirm Original Blood Vessels in 80 Million-Year-Old Fossil," NC State News, 12/1/15.

How Do Bats Land Upside Down?

Psalm 40:4
"Blessed is that man that maketh the LORD his trust, and respecteth not the proud, nor such as turn aside to lies."

Most bats, as you know, hang from their feet on the ceilings of caves. But not until recently did scientists know how bats switched from flying right-side-up to landing upside down in the space of a half second. Scientists assumed that bats used some kind of aerodynamic trick to pull this off. But the bats' mid-flight flip has almost nothing to do with air flow.

According to a new study in *PLOS Biology*, researchers took high-speed videos of bats as they landed upside down on a gauze net attached to the ceiling. The researchers were surprised to see the bats manipulating the inertia in their bodies to execute the flip just before landing. This is similar to the way pirouetting figure skaters speed up or slow down their rotation by pulling their arms closer to their body.

"It never would have occurred to me that aerodynamics would play such a small role in landing," said evolutionary biologist Sharon Swartz. "I always think of flight as a primarily aerodynamic phenomenon. Wings are aerodynamic organs, and landing seems so obviously to be a flight behavior."

Once again, we see that "obvious" answers can lead scientists astray. Evolutionary biologists in particular think it's obvious that animal behaviors are the result of evolution. They should start looking at solutions that aren't so obvious. If they did, they might learn more about the creatures they are studying as well as the One who created them.

Prayer: Heavenly Father, You have given Your creatures everything they need to live the life You planned for them, but You have done so much more for me by sending Your Son to die for my sins. In Jesus' Name. Amen.

Ref: N. Akpan, "Bats flip like Tony Hawk to land upside down," PBS Newshour, 11/16/15.

Who's Looking Out for You?

Genesis 28:15
"And, behold, I am with thee, and will keep thee in all places whither thou goest, and will bring thee again into this land; for I will not leave thee, until I have done that which I have spoken to thee of."

A cable TV news commentator often poses the question – "Who's looking out for you?" – implying that he's looking out for his viewers. Well, Christians know who's *really* looking out for us. That would be Jesus, the One who created us, saved us and watches out over us every hour of our lives. But in His wisdom, our Creator has also given us a number of animals to watch out for us in a more earthly sense. These creatures are known as animal sentinels.

Perhaps the best-known animal sentinel is the canary. As Wikipedia tells us, "The idea of placing a canary or other warm-blooded animal in a mine to detect carbon monoxide was first proposed by John Scott Haldane, in 1913." This is where the familiar phrase "canary in a coal mine" came from.

Over the years, man has turned to many other creatures to keep us safe. Honeybees, for example, have been used to test for air pollution. Bivalve mollusks help us do water-quality surveys. Pigeons are excellent at detecting lead in the atmosphere. Man's best friend has been used to provide early warning of lead-poisoning hazards in the home. Even cats have been helpful in determining household exposures to pesticides, cigarette smoke and other carcinogens.

Let us not forget to thank God for giving us these animal sentinels and for providing humans with the ingenuity to know how these animals can be used for our protection.

> ***Prayer: Heavenly Father, I just want to thank You for watching out for me, not only in this life but in the never-ending life to come. In Jesus' Name. Amen.***

Ref: Wikipedia entry on "animal sentinel."

"Then I Got Invited to a Bible Study"

Isaiah 55:11
"So shall my word be that goeth forth out of my mouth: it shall not return unto me void, but it shall accomplish that which I please, and it shall prosper in the thing whereto I sent it."

Dr. Gary Parker, a scientist with a doctorate in biology, used to love teaching his students about what he called "the fact of evolution." As he describes it in his book *Creation Facts of Life,* "For me, evolution was much more than just a scientific theory. It was a total world-and-life view, an alternate religion, a substitute for God."

He continued, "I didn't just believe evolution. I embraced it enthusiastically! And I taught it enthusiastically! I considered it one of my major missions as a science professor to help my students rid themselves of old, 'pre-scientific' superstitions, such as Christianity." Dr. Parker also points out that he was willing to let his students believe in whatever "god" they wanted to ... just as long as their religious belief did not keep them from believing in evolution.

His story takes a sudden turn, however, when he writes, "But then I got invited to a Bible study." For this distinguished scientist, everything changed. And for many years now, through his books and seminars, he has been using science and logic to teach that evolution is false. "And it's not just me," he adds. "Thousands of scientists are sharing the scientific evidences in God's world that encourage us to believe all the wonderful promises and principles in God's Word, the Bible."

It's truly amazing how adaptable even the scientific mind becomes when considering the complete body of evidence rather than conforming to unproven hypotheses like evolution.

Prayer: Father, I know that the simple act of inviting someone to a Bible study can have a profound impact on that person's life and on the lives of many others. Use me, Lord, to reach out to others. Amen.

Ref: Dr. Gary Parker, *Creation Facts of Life,* pp. 11-12 (Master Books, 15th printing, 2013).

Ota Benga: Man or Monkey?

John 12:26
"If any man serve me, let him follow me; and where I am, there shall also my servant be: if any man serve me, him will my Father honour."

In his review of Dennis Sewell's book *The Political Gene*, creation scientist and author Dr. Jerry Bergman mentions one of the most deplorable events in evolutionary history – the story of Ota Benga. A pygmy from the Congo, Benga attracted a great deal of attention when he was put on display at the Bronx Zoo. He eventually ended his life by firing a bullet into his heart in 1916.

Ota Benga and an orangutan of similar height were placed inside a locked cage. Quite naturally, the display had people asking if Ota Benga was a monkey or a man. The zookeeper answered that he was a transitional form between man and monkey – the missing link. Sewell's book points out that evolutionists defended the display by noting that evolution is taught in the school textbooks and is "no more debatable than the multiplication table."

African American clergymen complained to zoo officials about the exhibit, saying, "The Darwinian theory is absolutely opposed to Christianity, and a public demonstration in its favor should not be permitted." But the *New York Times* disagreed, saying: "We do not quite understand all the emotion which others are expressing in the matter. It is absurd to … moan over the imagined humiliation and degradation Benga is suffering. The pygmies ... are very low in the human scale…"

The tragic story of Ota Benga clearly reveals the stark contrast between those who follow Christ and those who follow Darwin. Whose side are you on?

Prayer: Heavenly Father, I pray that You will open the eyes of many so they will turn away from Darwin's false teachings and turn to You for eternal life. In Jesus' Name. Amen.

Ref: Dr. Jerry Bergman, "Darwin is the universal acid that affects everything," *Journal of Creation*, Vol. 25(1), 2011, pp. 19-21.

The Fermi Paradox

Deuteronomy 10:14
"Behold, the heaven and the heaven of heavens is the LORD'S thy God, the earth also, with all that therein is."

Where is everybody?! That's what physicist Enrico Fermi wanted to know. After all, if life evolved on Earth, he thought, certainly it must have evolved on countless other planets. So in 1950, he put forth what is called the "Fermi paradox" – a challenge to other scientists to explain why extraterrestrial life has not yet been found.

One of the more popular explanations is that the universe is so big, we just haven't found them yet. But now there are scientists who think aliens may have already found us! They are hiding from us to protect themselves, according to Adrian Kent of the Perimeter Institute in Ontario, Canada. Aliens, he suggests, do not want to attract the attention of more advanced species from other planets who might want to harm them.

But let's think about that. If aliens have developed the technology to travel through the vastness of space to a planet that only managed to put a man on its moon around fifty years ago, wouldn't they be far more advanced than us and have nothing to fear?

Many other scientists have attempted to solve the Fermi paradox. Some of their solutions are very imaginative. But here is one solution that I believe agrees with both science and the Bible. The reason why man has not found other beings elsewhere in our vast universe is because the Earth is the only planet created by God to nurture human beings, who were made in the image of God Himself.

Prayer: Heavenly Father, I pray that scientists would turn their attention to learning about the universe that You created – just like it really is and not like they want or imagine it to be. In Jesus' Name. Amen.

Ref: "Is ET hiding?" *Creation* magazine, Jan-Mar 2012, p. 9.

So Where Is Everybody?

Jeremiah 10:12
"He hath made the earth by his power, he hath established the world by his wisdom, and hath stretched out the heavens by his discretion."

On our previous broadcast, we told you about the Fermi paradox. In case you missed it, the Fermi paradox is the attempt by scientists to account for the fact that we have not come in contact with extraterrestrial life. This is a very real problem for evolutionists because they believe that if life evolved here, evolution must have occurred on countless other planets. So why haven't we found anybody out there? Here are a few attempts to solve the Fermi paradox.

Because intelligent civilizations are too far apart in space-time, say some. Because it is too expensive to travel throughout the galaxy, say others. Still other scientists believe it's because civilizations broadcast detectable radio signals only for a brief period of time. Or how about this one – that alien life forms are so different from us, we cannot recognize their attempts to communicate with us. And die-hard evolutionists suggest that intelligent extraterrestrial life does exist, but they don't contact us because they don't want to interfere with our natural evolution and development.

Creation Moments believes that all of these explanations fall short because the Fermi paradox is based on the faulty premises that evolution is true and that life must have evolved elsewhere. As we said last time, we believe the real reason why mankind hasn't found intelligent life out there is because God created life on only one planet – right here on Earth.

Prayer: Heavenly Father, I pray that many of those who are looking for extraterrestrials will come to realize that our planet was already visited by Someone who is not of this Earth. In Jesus' Name. Amen.

Ref: Wikipedia entry on "Fermi paradox."

Evolutionists Still Clueless on Origin of Life

Job 32:3
"Also against his three friends was his wrath kindled, because they had found no answer, and yet had condemned Job."

How good are evolutionists at explaining the origin of life? Not very! Listen to what some have written and said. First up is professor Paul Davies from the University of Arizona. In his book *The Fifth Miracle: The Search for the Origin and Meaning of Life*, he wrote, "Obviously, Darwinian evolution can operate only if life of some sort already exists." He continued: "Darwinism can offer absolutely no help in explaining that all-important first step: the origin of life."

Now consider what evolutionist science writer Gordy Slack wrote in *The Scientist*: "Evolution should be able to explain, in theory at least, all the way back to the very first organism that could replicate itself through biological or chemical processes ... and what came before it.... Right now, we are nowhere close."

Here's another honest statement – this one from Simon Morris, professor of evolutionary biology at Cambridge University: "Despite decades of experimentation, with accompanying shouts of 'breakthrough' or 'almost there,' we are still paddling on the edges of an ocean of ignorance."

I close today's broadcast with this marvelous quote from Henry Lipson, professor of physics at the University of Manchester: "I think we must admit that the only acceptable explanation is *creation*. I know that this is anathema to physicists, as indeed it is to me, but we must not reject a theory that we do not like if the experimental evidence supports it."

Prayer: Heavenly Father, I pray that when evolutionists run into walls that the tools of science can't break through, they will turn to biblical creation for the truth. Amen.

Ref: Sources of all quotes can be found in *Evolutionists Say the Oddest Things*, Lita Cosner, editor, pp. 43-46 (Creation Book Publishers, 2015).

Creationist Astrophysicist

Psalm 8:3
"When I consider thy heavens, the work of thy fingers, the moon and the stars, which thou hast ordained;"

Evolutionists would be foolish to question creation scientist Dr. Markus Blietz's credentials as a legitimate scientist. Dr. Blietz earned his Ph.D. in astrophysics at the Max Planck Institute for Extraterrestrial Physics and now works in the semiconductor industry in charge of helping researchers develop new ideas.

In an interview that was published in the book *Busting Myths,* he talks about black holes, comets, distant starlight, and most important, how he came to become a Christian and a creationist. "I read the gospel according to Matthew," he said. "Almost immediately I understood that Jesus was a real, historical person, that He came to fulfill a mission and that I needed Him urgently."

He also spoke out about evolution and world views. "Only belief in Jesus Christ can open our eyes and give us the correct view of our world," he said, adding that "before I was a Christian, I never felt really content with the evolutionary world view, which I had adopted. It produced too many contradictions and left open too many questions. Only the truth in the Word of God is able to give a full, comprehensive answer to our basic questions of life and death."

Dr. Blietz also had these words of advice for young people: "Study science to be able to better serve God." Creation Moments thinks this is great advice, not just for young people but for everyone.

Prayer: Heavenly Father, thank You for raising up scientists who know biblical truth. I pray that You will encourage and protect them as they take a stand for Your truth among their unbelieving colleagues. Amen.

Ref: "Galaxies, Black Holes, and Creation," Jonathan Sarfati and Markus Blietz, *Busting Myths*, pp. 11-15 (Creation Book Publishers, 2015).

Are Biblical Creationists a Stumbling Block?

Mark 10:5-6
"And Jesus answered and said unto them, For the hardness of your heart he wrote you this precept. But from the beginning of the creation God made them male and female."

In his excellent article in *Creation* magazine "Jesus on the Age of the Earth," Carl Wieland reports that the standard secular timeline of billions of years for the age of the universe is "accepted by most people in the evangelical Christian world." He went on to write, "Some would even say that to even dispute billions of years is to place an unnecessary stumbling block in the way of any scientifically-minded potential converts."

I can't count the number of times Creation Moments has been attacked by Christian leaders and laypeople … and even pastors! No, young earth creationists are not the stumbling block. We are simply standing up for biblical authority in a world that is rejecting it. As Wieland writes, many authors of the Bible – even Jesus Himself – make it clear that people were on the Earth at the beginning of creation and not millions of years later.

How tragic, writes Wieland, that so many Christian leaders have been "bluffed and intimidated into assuming that secular interpretations of the evidence should dictate their understanding of God's Word." Ironically, this rejection of biblical authority comes "at a point in history when there are more scientific reasons than ever to confirm the utter rationality of trusting the Bible…"

Yes, you can and must trust the Bible. After all, you don't want to become a stumbling block to the gospel by rejecting what the Bible clearly teaches.

> *Prayer: Father, I pray that You will keep me from being a stumbling block in the way of unbelievers. I believe that the best way to present the gospel is by teaching what the Bible itself teaches. In Jesus' Name. Amen.*

Ref: Carl Wieland, "Jesus on the Age of the Earth," *Creation* 34(2) 2012, pp. 51-53.

The Goldilocks Zone

Genesis 8:22
"While the earth remaineth, seedtime and harvest, and cold and heat, and summer and winter, and day and night shall not cease."

I'm sure you remember the fairytale of Goldilocks and the Three Bears. Goldilocks tries to eat papa bear's porridge but finds it too hot. She then turns to a second bowl, and it's too cold. Finally she discovers a third bowl of porridge and finds it to be just right, so she eats it all up.

Since the 1970s, astronomers have described our own planet as being situated in what they call a cosmic Goldilocks Zone that is just right for life to exist. It is neither too hot nor too cold, neither too far away from the sun nor too close. Years ago, however, NASA noted that the Goldilocks Zone is much larger than they ever expected.

"Scientists have found microbes in nuclear reactors, microbes that love acid, [and] microbes that swim in boiling-hot water," wrote NASA. They also observed, "Whole ecosystems have been discovered around deep sea vents where sunlight never reaches and the emerging vent-water is hot enough to melt lead."

But now, scientists are once again rethinking their position. *Nature* magazine reported that "tidal heating shrinks the 'goldilocks zone' and concluded that this "overlooked factor suggests fewer habitable planets than thought."

With each new discovery in recent years, the size of the Goldilocks Zone is shrinking. Perhaps one day scientists will realize that only one planet in the universe occupies the Goldilocks Zone ... the planet we call home!

Prayer: Heavenly Father, thank You for placing the Earth in the exact spot of the universe which is suitable for life. In Jesus' Name. Amen.

Ref: "The Goldilocks Zone," NASA Headline News, 10/2/03. R.A. Lovett, "Tidal heating shrinks the 'goldilocks zone'," *Nature*, 5/8/12.

You Are Mostly Bacteria

Leviticus 17:11
"For the life of the flesh is in the blood: and I have given it to you upon the altar to make an atonement for your souls: for it is the blood that maketh an atonement for the soul."

You may want to be sitting down as you listen to today's broadcast. Why? Because scientists tell us there are far more bacterial cells in your body than actual human cells. According to researchers at the Weizmann Institute of Science in Israel and the Hospital for Sick Children in Toronto, an average adult weighing 155 pounds consists of about 40 trillion bacteria and only 30 trillion human cells.

This estimate could be off by as much as 25 percent, the researchers admit. The actual number of bacteria could be as high as 50 trillion. Nobody knows for sure because no one has ever counted them. Previously, scientists guesstimated that bacteria could outnumber human cells by ten to one or even a hundred to one. In a 2014 issue of *Microbe* magazine, however, molecular biologist Judah Rosner called the ten-to-one ratio a "fake fact," saying "everybody likes a nice, round number."

Now, when you consider that viruses, fungi and other microbes far outnumber the bacteria in your body, you might start feeling that there's very little of "you" left! But remember these two things. Human cells have a lot more mass than bacteria. And 85 percent of your human cells are red blood cells – the very same kind of cells the Son of God shed on the cross to rescue you from the penalty of sin.

And that is most definitely *not* a fake fact!

Prayer: Heavenly Father, thank You for sending Jesus to lay down His life as a sacrifice so I can look forward to eternal life! In Jesus' Name. Amen.

Ref: T. H. Saey, "Human Body Not Overrun by Bacteria," *ScienceNews*, 2/6/16, p. 6.

Night of the Living Zombie Ant!

Romans 7:25b
"So then with the mind I myself serve the law of God; but with the flesh the law of sin."

Zombies are all the rage today in movies and television series. On today's broadcast, however, we will be looking at a real-life zombie, produced not by Hollywood but by a fungus. The *Ophiocordyceps* fungus literally turns carpenter ants into zombie slaves.

By chemically manipulating the ant's brain, the fungus causes the ant to leave the colony's comfortable nest in the tree canopy and move to the underbrush closer to the ground. Penn State entomologist David Hughes explains that "infected ants behave as zombies and display ... behaviors of random rather than directional walking." By walking in such an erratic manner, the fungus prevents the ant from climbing back to the canopy.

According to writer Chad Meeks, "the fungus keeps the ant wandering about 10 inches above the forest floor – and then, right at solar noon, the ant bites as hard as it can into a leaf" – something these ants would never dream of doing. "Then the ant locks its jaw and dies there." Two or three days later, "a stalk erupts from the dead ant's head. It soon begins shooting out new fungal spores, which will be picked up by more carpenter ants. And the cycle begins again."

The sin nature that corrupts and controls us comes not from a fungus but from the first Adam. Thankfully, though, the last Adam – Jesus Christ – died on the cross to free us from the control and the curse of sin.

Prayer: Heavenly Father, forgive me for letting my sin nature draw me away from You and Your ways. Help me to hate sin as much as You do! In Jesus' Name. Amen.

Ref: "I Want to Be a Zombie Ant," Chad Meeks, adjunct professor at Southwestern Baptist Theological Seminary. *The Behemoth*, published by the editors of *Christianity Today*, 11/12/15.

From Earth to Mars in Three Days

1 Corinthians 15:52
"In a moment, in the twinkling of an eye, at the last trump: for the trumpet shall sound, and the dead shall be raised incorruptible, and we shall be changed."

Even if you haven't seen the movie *The Martian*, you know it takes several months for a spacecraft from Earth to reach the red planet. But if NASA has anything to say about it, that trip might someday take only a matter of days.

Dr. Philip Lubin, physics professor at the University of California, Santa Barbara, suggests using a technology called photonic propulsion. He thinks it could reduce the travel time from Earth to Mars to just three days for a spacecraft weighing 220 pounds.

If you're like me, you're probably thinking this is the sort of thing you would expect to find in science-fiction films. But scientists at NASA are dead serious about Directed Propulsion for Interstellar Exploration – or DEEP IN for short. According to an article at *Universe Today*, "Directed Energy Propulsion differs from rocket technology in a fundamental way: the propulsion system stays at home, and the craft doesn't carry any fuel or propellant. Instead, the craft would carry a system of reflectors, which would be struck with an aimed stream of photons, propelling the craft forward."

The article continues, "And if that's not tantalizing enough, the system can also be used ... to *detect other technological civilizations.*"

But of course! Hasn't this been one of NASA's primary missions for many years now? NASA wants to find other civilizations to show that the Earth is not the unique planet God created it to be. Now, *that* is science-fiction!

Prayer: Heavenly Father, I can't wait for the day to arrive when I will be raised incorruptible in the twinkling of an eye – at the speed of light! In Jesus' Name. Amen.

Ref: E. Gough, "NASA thinks there's a way to get to Mars in three days," *Universe Today*, 2/24/16.

Has Evolution Produced Anything of Practical Value?

Job 38:4
"Where wast thou when I laid the foundations of the earth? declare, if thou hast understanding."

Many of our listeners are familiar with the name of Dr. Don Batten – a plant scientist and creationist who authored countless books, articles and booklets on science and creation. Recently we came across an interview where Dr. Batten was asked to respond to the claim being made by evolutionists that science would be impossible without evolution.

Listen closely to what he said: "I don't know anything of practical value in science that has come out of evolutionary thinking. In fact," he added, "evolutionary daydreaming has given rise to many dead ends." As an example, he mentioned "the inappropriate treatment of back pain – trying to make our backs more like those of our supposed ape ancestors – which," he added, "actually makes back problems worse."

But Dr. Batten wasn't finished: "Then there are the false notions that some organs are useless leftovers of evolution, or that DNA that we don't understand is 'junk'. Such ideas," he said, "impede scientific progress, as they influence scientists not to bother investigating their function."

He also mentioned the many frauds committed in the name of evolution – like Haeckel's embryo drawings, Piltdown Man and the feathered dinosaur *Archaeoraptor*. All of these and many more hoaxes have stood squarely in the way of scientific progress.

So, young people, if you love science, do what Dr. Batten has done – become a creation scientist and accomplish great things!

Prayer: Father, though I know that science has become hostile to people of faith, I pray that You will make a way for young Christians to serve You as scientists. Amen.

Ref: "Harvesting Real Fruit," *Busting Myths: 30 Ph.D. scientists who believe the Bible and its account of origins*, edited by Jonathan Sarfati and Gary Bates, p. 64 (Creation Book Publishers, 2015).

God's Special Gift to Salmon

Acts 26:18a
"To open their eyes, and to turn them from darkness to light, and from the power of Satan unto God, that they may receive forgiveness of sins..."

Creation Moments has aired hundreds of programs over the years dealing with gifts God has given to His creatures so they could survive in the environment where He has placed them. But what is a poor animal going to do when it lives most of its life in one environment and then moves to a totally different environment?

Today I'm going to tell you about a special gift that God has given to salmon. The purpose of this gift will become clear to you when you consider that salmon live most of their lives in ocean waters, where the light is in the blue-green end of the spectrum. Later on, salmon travel to streams and inland waters to spawn, where the light is primarily in the red and infrared end of the spectrum. Scientists have long wondered how salmon are able to make the switch so they can see well in two entirely different lighting conditions.

Scientists have now discovered an enzyme in salmon that actually switches the fish's visual systems. What this enzyme does is convert vitamin A1 to vitamin A2, allowing salmon to see perfectly in red and infrared light.

Like salmon, humans also need a new way of seeing when it comes to spiritual matters. The eyes we are born with just aren't equipped to see Jesus in a favorable light. We need eyes of faith that God alone can provide!

Prayer: Heavenly Father, I pray that You will give unbelievers eyes of faith that can see Your Son and follow Him on the path that leads to life everlasting. In Jesus' Name. Amen.

Ref: "Ancient gene triggers IR vision enhancement," *Photonics Spectra,* January 2016, p. 106.

Caution: Fuzzy Words Ahead!

Psalm 119:160
"Thy word is true from the beginning: and every one of thy righteous judgments endureth for ever."

Science textbooks, articles and documentaries about evolution are saturated with what creationist Mike Riddle calls "fuzzy words." These are words that evolutionists typically use when they have no observable evidence to support their claims.

With that as a backdrop, today's broadcast will train you to spot fuzzy words and phrases. In a recent issue of *ScienceNews* magazine, for example, we find the first fuzzy words in the article's title: "Bubbles may have sheltered early life." Did you spot it? Yes, it's the phrase "may have."

In the second paragraph, we read: "Such a snug hideout could have shielded microbes from ultraviolet radiation." Right – the words "could have." In the next paragraph, a geologist is quoted as saying that the work is "very plausible" – another way of getting you to accept the scientist's wishful thinking. In the remaining paragraphs, we find two more "could haves", a "may have", a "perhaps", a "might have" and two instances of "appeared to be".

The writer then concludes with these words: "If microbes survived in these pockets on early Earth, they could potentially have done so on other planets such as Mars." I hope that the words "if" and "potentially" set off the critical thinking alarm bells in your brain.

I'm so glad that the Bible avoids fuzzy words and phrases, aren't you? The Bible doesn't try to trick or deceive us. Its words are true because its Author is Truth and His words endure forever.

Prayer: Heavenly Father, You say what You mean and mean what You say in Your inspired Scriptures! If You had created the heavens and Earth in a different way, I know You would have written it that way! In Jesus' Name. Amen.

Ref: M. Rosen, "Bubbles may have sheltered early life," *ScienceNews*, 2/6/16, p. 12. Mike Riddle, "Critical Thinking: Examining Their Words," 12/31/15, Creation Training Initiative.

Can You Wiggle Your Ears?

Proverbs 1:22
"How long, ye simple ones, will ye love simplicity? and the scorners delight in their scorning, and fools hate knowledge?"

Evolutionists tell us that the muscles which control movement of the ears – the auriculars – are vestigial. That is, they might have been useful to humans in our evolutionary past, but other than allowing some people to wiggle their ears, they have no purpose today. That's why a new review of research on these muscles is causing so much excitement among evolutionists. "According to intelligent design and creationism," the researcher writes, "our body was designed by a being with perfect intelligence. Here's something in our brain that's completely useless, so why would a being of perfect intelligence put it there?"

Biologist and creationist anatomist Dr. David Menton fired back with something for evolutionists to think about. This former professor at Washington University School of Medicine wrote: "One of the problems with the whole concept of vestigial or functionless muscles is the well-known fact that unused muscles quickly degenerate."

He went on: "It is unlikely that any muscle that was virtually unused for the lifetime of an individual (to say nothing of generations of individuals over millions of years) would remain as healthy muscle tissue. It seems overwhelmingly likely that any muscle in the body that actually exists in the present, serves some function."

Creation Moments has no doubt that Dr. Menton is correct in saying that the auricular muscles have a purpose. We do doubt, however, that evolutionists will make a serious attempt to discover what that purpose might be!

Prayer: Father, I know that scorners delight in their scorning and fools hate knowledge. Even so, I pray You will save many through Your divine grace. Amen.

Ref: Stephanie Pappas, "No Purpose for Vestigial Ear-Wiggling Reflex," *LiveScience*, 10/22/15. David Menton, "The plantaris and the question of vestigial muscles in man," Countering the Critics, CEN Technical Journal 14(2) 2000, p. 50.

Evolutionary Hype!

Job 17:5
"He that speaketh flattery to his friends, even the eyes of his children shall fail."

In a surprisingly candid article at the pro-evolution website *ScienceDaily*, one scientist admits that most fossil finds include a generous amount of hype.

Dr. James Tarver from the University of Bristol noted that "human fossils are very rare, and they are costly to recover because of the time involved and their often remote locations." Because of this, he said, "scientists may be pushed by their sponsors, or by news reporters, to exaggerate the importance of their new find and make claims that 'this new species completely changes our understanding'."

In other words, don't believe the hype you hear from scientists. The article doesn't mean, though, that Dr. Tarver is backing away from evolution. Far from it. The *ScienceDaily* article goes on to say that the study – and I quote – "suggests most fossil discoveries do not make a huge difference, confirming, not contradicting our understanding of evolutionary history."

Such a statement clearly demonstrates that scientists interpret fossils within their evolutionary framework, finding what they expect to find.

So what do they do when they find a fossil that flat-out contradicts Darwinism? Sometimes they come up with imaginative explanations to make the fossil fit their beliefs. Or even worse, they sweep the evidence under the rug – like what they've done with dinosaur bones containing soft tissues and blood cells.

The lesson to be learned – don't fall for evolutionary hype ... and never resort to using hype yourself. The truth can stand on its own.

Prayer: Heavenly Father, help Your people expose the hype behind the latest science headlines so that the world will see the emptiness of evolution. In Jesus' Name. Amen.

Ref: University of Bristol. "Evolution rewritten, again and again," *ScienceDaily*, 9/1/10.

Windows of Heaven

Isaiah 54:12
"And I will make thy windows of agates, and thy gates of carbuncles, and all thy borders of pleasant stones."

If the streets of heaven are made of pure gold and its twelve gates are pearl, what are the windows of heaven made of? The King James version of the Bible tells us: "And I will make thy windows of agates."

What is an agate? According to geologist and agate collector Bill Kitchens, "Agates are treasures of God's grace that, through natural processes, fill voids in rocks around us for people to find, wonder at, and use."

How fitting a description – not in scientific terms but in the emotional impact agates can have on us. Kitchens has put together a website – GodMadeAgates.com – that includes an impressive collection of photos he has taken, showing the incredible beauty and variety of agates.

Now, why would a geologist spend so much time photographing agates? As Kitchens tells it, "For as long as I can remember, I have been fascinated with agates, and have learned lots about them, though even now I'm frequently stumped when I look deeply into these complex creations."

He admits he is also perplexed by what he sees in his fellow human beings. But there are two things he is absolutely sure about: "The things I know best and am most certain of are the grace of God, which has upheld me all these years, and the trustworthiness of His word, which, as the years go by, I have come more and more to appreciate."

> ***Prayer: Heavenly Father, thank You for the wonders of Your creation – even those that are the result of natural processes. After all, Your Son was responsible for putting those natural processes in place! In Jesus' Name. Amen.***

Ref: Bill Kitchens, "Speaking of Agates and God, and Man," http://godmadeagates.com.

Make Noise, Not War

Ezra 3:13
"So that the people could not discern the noise of the shout of joy from the noise of the weeping of the people: for the people shouted with a loud shout, and the noise was heard afar off."

On today's program I will give you one guess at what howler monkeys are especially good at. That's right – they howl. Or perhaps I should say, they shout. According to the *Guinness Book of World Records*, their shouts can be heard clearly up to three miles away, making them the loudest land animal on record.

Now, why did God give them this ability to shout so loudly to other howler monkeys? Howlers live in groups of about twenty individuals. And since they would rather make noise than war, when one group of monkeys is about to encroach upon the territory of a different group, the monkeys make their presence known by howling. This enables them to avoid violent confrontations.

Males are a great deal louder than females. That's because they have a hollow bone near their vocal cords – the hyoid – which amplifies the sound by acting like the body of a drum. And according to the book *1000 Wonders of Nature,* males also have a thick neck and drooping double chin that "acts as a resonating chamber to amplify the sound."

God wants His people to raise our voices, too, but with songs of praise. As our Creator and Savior, He alone is worthy of our praise. But He also wants us to use our voices to present the gospel to unbelievers so that they, too, might trust in Jesus for the forgiveness of their sins.

> *Prayer: Heavenly Father, I pray that You will bless the words I speak to others so that they might come to know You as their Savior. In Jesus' Name. Amen.*

Ref: "The long-distance call of the howler monkey," *1000 Wonders of Nature*, p. 204 (Reader's Digest, London).

Human Hand Is Evidence of a Creator

Psalm 2:2-4
"The kings of the earth set themselves, and the rulers take counsel together, against the LORD, and against his anointed, saying, Let us break their bands asunder, and cast away their cords from us. He that sitteth in the heavens shall laugh: the Lord shall have them in derision."

Is the human hand evidence of a Creator? In a paper which appeared in the science publication PLOS ONE, a team of four scientists claimed that the link between muscles and hand movements is the product of "proper design by the Creator." The paper goes on to say that human hand coordination "should indicate the mystery of the Creator's invention," and concludes by again claiming the mechanical architecture of the hand is the result of "proper design by the Creator."

How shocking for such a paper to appear in a pro-evolution science journal! After all, such publications do not publish articles that even hint at the existence of a Creator. Well, as you would expect, the scientific community was quick to call for PLOS ONE to retract the article, and that's exactly what they did. The publisher also apologized to readers who were threatening to boycott the publication.

A situation like this only underscores the fact that science – led for the most part by atheists – has become an exclusive club that scientists like Sir Isaac Newton, James Clerk Maxwell and Louis Pasteur would not be permitted to join. Indeed, even present-day scientists like MRI inventor Dr. Raymond Damadian would be told to take a hike.

Try as they might to remove God from the picture, however, the fact remains that He who sits in the heavens is laughing at them in derision. He, not they, will have the last laugh.

Prayer: Father, because evolution keeps people on the broad road that leads to destruction, I pray You will give me many opportunities to share the gospel. Amen.

Ref: "Scientific paper which says the human hand was designed by a 'Creator' sparks controversy," *Independent*, 3/3/16.

The Bird You Must Hear to Believe

1 Chronicles 16:23
"Sing unto the LORD, all the earth; shew forth from day to day his salvation."

One of God's most amazing and amusing creatures is the Superb Lyrebird, a bird found in the rainforests of Victoria, New South Wales and southeast Queensland, in Tasmania and along the east coast of Australia. After British naturalist and filmmaker David Attenborough turned his cameras on them, the lyrebird achieved worldwide fame for its unique ability to mimic the calls of virtually anything they hear.

Citing the *Handbook of the Birds of the World*, Wikipedia noted that lyrebirds have been recorded mimicking human sounds such as a mill whistle, a cross-cut saw, chainsaws, car engines and car alarms. The birds are also able to duplicate the sound of fire alarms, rifle-shots, camera shutters, dogs barking, crying babies, music, mobile phone ring tones, and even the human voice.

Though lyrebirds sing all year long, the males save their best and most intense songs during the peak of breeding season from June through August. But you can hear their marvelous songs anytime you like. Simply do a web search on lyrebirds or look for lyrebird videos on YouTube.

Unlike the larynx of humans, the vocal organ of birds is called the syrinx. And the lyrebird's syrinx has more muscles than any other songbird. Still, evolutionists have no words to satisfactorily explain how such vocal abilities came about through natural means. But creationists have no difficulty in giving praise to God for this magnificent creature!

Prayer: Heavenly Father, when I hear the amazing songs of the lyrebird, I want to raise my own voice in praise for all the magnificent flying creatures made on the fifth day of creation! Amen.

Ref: Wikipedia entry on "Lyrebird."

Eyewitness to a Volcano's Birth

Judges 5:5
"The mountains melted from before the LORD, even that Sinai from before the LORD God of Israel."

How long does it take for a volcano to form? Scientists who reject the flood of Noah's time insist that it takes many thousands or millions of years. And yet, we know from eyewitness testimony that at least one volcano – Mount Paricutin in Mexico – grew to over 300 feet in height within a week of its birth.

The birth of Mount Paricutin took place in 1943, when a Mexican farmer, Dionisio Pulido, along with his son and wife, were working in a field. He was frightened when he saw a fissure open in the ground and the surrounding soil starting to bulge upwards. Soon the air was filled with dark smoke and gas smelling like rotten eggs.

Before the volcano became extinct nine years later, it had reached a height of 1,390 feet – tall enough to be seen from miles away. Since 1952, the volcano has been attracting tourists and climbers from all over the world. Now, if you were to ask tourists who hadn't listened to today's program how long the volcano has been there, they would say thousands or millions of years.

People generally say millions of years when asked about the age of the Grand Canyon. But they are basing their answers on assumptions made by geologists who don't take a worldwide flood into account. Far better than assumptions is the geological history of the world we find in the Bible.

Prayer: Heavenly Father, though I know I will be mocked, I believe what the Bible tells me about the six-day creation and the worldwide flood. After all, You were there so You ought to know! In Jesus' Name. Amen.

Ref: Jonathan O'Brien, "Paricutin," Creation 35(1) 2013, pp. 32-33.

Bigmouth!

Jonah 1:17
"Now the LORD had prepared a great fish to swallow up Jonah. And Jonah was in the belly of the fish three days and three nights."

In 1976, a naval research vessel working in the deep waters off Hawaii discovered a large fish that was entangled in its anchor. When the sailors hauled the fish aboard, they saw something no one had ever seen before. This fish – now considered to be one of the most important marine discoveries of the twentieth century – was over fourteen feet in length and weighed 1,650 pounds.

What they had discovered was a shark with a very, very, VERY big mouth! That's why this fish is now commonly known as the megamouth shark.

With its enormous mouth, you might expect it to feed on large prey, but its gaping mouth is actually filled with very tiny teeth. Megamouth is a filter feeder. When jellyfish, shrimp and other small invertebrates are nearby, the shark simply opens its mouth, sucks in a large amount of water and filters out its lunch.

Megamouth sharks swim in a vertical position, with their mouth pointing straight up towards the surface. They also migrate vertically. That is, they spend their daylight hours submerged to a depth of about 500 feet. Then at night they migrate vertically to approximately 50 feet.

The megamouth shark reminds us of an event recorded in the Bible book of Jonah. No, we're not talking about the big fish that swallowed the disobedient prophet. We're talking about Jonah himself. With his constant complaining against God, what a bigmouth he turned out to be!

Prayer: Heavenly Father, keep me from complaining about my circumstances. Remind me to praise You in the midst of my difficulties. In Jesus' Name. Amen.

Ref: Tim Flannery and Peter Schouten, "Megamouth," *Astonishing Animals,* pp. 110-111 (Atlantic Monthly Press, 2004). Also, Wikipedia entry on "Megamouth shark."

How Cats Really Drink Milk

John 4:10
"Jesus answered and said unto her, If thou knewest the gift of God, and who it is that saith to thee, Give me to drink; thou wouldest have asked of him, and he would have given thee living water."

Today's program is dedicated to all of you cat lovers out there. Now, you probably remember being told that your precious pet laps up milk by curling his tongue backwards, and he then uses his curved tongue to ladle the liquid into his mouth. Well, I'm sure you'll be happy to know that your cat is a lot more talented than that.

Using high-speed digital video, physicists have now discovered that only the tip of your cat's tongue comes into contact with the milk. As your cat retracts his tongue, a small column of milk follows it into the air. Now, when the inertia of the milk going upwards matches the force of gravity pulling the milk back down, the liquid hovers in mid-air for a split second. And it's at this precise moment when your cat closes his mouth over the milk.

And if this wasn't impressive enough, your cat performs this maneuver four times per second! According to a Virginia Tech researcher, "We were surprised and amused by the beauty of the fluid mechanics involved in this system."

Surprised? Creationists aren't surprised. After all, the God who created fluid mechanics is the same God who created cats, and He knew exactly what abilities cats would need to quench their thirst. Thankfully, He also knows how to quench man's thirst for eternal life, and He freely dispenses the "living water" of the gospel.

Prayer: Heavenly Father, I pray that many friends and family members will see that the only way to satisfy their thirst for eternal life is the living water provided freely by Your Son, Jesus Christ. Amen.

Ref: David Catchpoole, Ph.D., "How Cats Drink Milk," Creation 35(1) 2013, p. 56.

One of Evolution's Best-Kept Secrets

Psalm 17:12-13
"Like as a lion that is greedy of his prey, and as it were a young lion lurking in secret places. Arise, O LORD, disappoint him, cast him down: deliver my soul from the wicked, which is thy sword:"

Though the title of today's program is "One of Evolution's Best-Kept Secrets," Creation Moments could bring you hundreds of broadcasts with the same title. While evolutionists are filling science textbooks, Hollywood films, science magazines and natural history museums with their favorite evidences of evolution, they routinely fail to mention the evidences that call evolution into question.

For example, evolutionists tell us that dinosaurs – after millions of years of gradual change – evolved into birds. But they don't mention that fossils of many modern birds have been found in the very same rock layers where dinosaurs are found!

As Dr. Carl Werner points out in his book and DVD *Living Fossils*, "Every time you see a T-rex or a Triceratops in a museum display, you should also see ducks, loons, flamingos or some of these other modern birds that have been found in the same rock layers as these dinosaurs, but this is not the case."

To see if this was an innocent omission or deliberate deception, Dr. Werner traveled to sixty natural history museums and ten dinosaur dig sights in seven different countries. His interviews with paleontologists revealed that they were well aware of the modern birds living alongside dinosaurs. And yet only one museum gave any hint that dinosaurs and modern birds lived at the same time.

What else aren't you being told about evolution? Keep on listening to Creation Moments because we will expose more of evolution's best-kept secrets on future broadcasts.

Prayer: Father, if evolution were true, evolutionists wouldn't have anything to hide. I pray that many of evolution's darkest secrets will come to light! Amen.

Ref: Dr. Don Batten, "Modern birds found with dinosaurs: Are museums misleading the public?" Creation 34(3), 2012.

What Really Wiped Out the Dinosaurs?

Psalm 33:4
"For the word of the LORD is right; and all his works are done in truth."

What caused dinosaurs to virtually disappear from the face of the earth? Was it an asteroid, as scientists now tell us? According to Dr. Henry Gee, science writer for *Nature*, "Nobody will ever know what caused the extinction of the dinosaurs because we weren't there to watch it happen."

Nevertheless, this hasn't stopped scientists from acting as if they were eyewitnesses. In his book *In Search of Deep Time*, Dr. Gee includes a list of dinosaur extinction explanations that paleontologist Mike Benton culled from scientific literature. Here's a small sampling:

Dinosaurs died out "because the climate became too hot, too cold, too wet, too dry, or a combination of the above; because their eggshells became too thin ... or too thick"; because of hay fever; because of indigestion from eating flowering plants. And here's my favorite: "Because, having been the dominant life-forms for 150 million years, the dinosaurs just got bored."

Let me suggest another explanation. Since dinosaur fossils are found in sedimentary rock layers all over the world, it's plain to see that most dinosaurs perished in the worldwide flood of Noah's day. Many dinosaurs that were aboard Noah's ark probably died out soon after the flood because they could not adjust to the post-flood climate. Many others were hunted by humans who were now permitted to eat meat.

But as we've shown on previous broadcasts, some dinosaurs survived and were seen well into the middle ages.

Prayer: Heavenly Father, I pray that You will prepare me to answer questions from unbelievers and evolutionists. Most of all, make me bold in proclaiming the gospel of grace. In Jesus' Name. Amen.

Ref: "Dinosaur demise," *Evolutionists Say the Oddest Things,* Lita Cosner, editor, p. 98 (Creation Book Publishers, 2015). Henry Gee, *In Search of Deep Time* (The Free Press, 1999).

Why Do We Sigh?

Psalm 150:6
"Let every thing that hath breath praise the LORD. Praise ye the LORD."

As we all know, coughing is a reflex that clears our breathing passages from secretions, irritants, foreign particles and even microbes that might make us sick. We also know that sneezing is a reflex that cleanses our nasal cavity by forcefully expelling foreign particles and irritants. But why do we sigh? Is it just because we're exasperated, surprised … or bored?

Well, here's a surprise for you. Though you aren't even aware of it, you take a deep breath – that is, you sigh – about twelve times every hour. As researchers have discovered, we sigh about every five minutes to keep our lungs functioning properly.

In fact, if we didn't sigh, we would die. Researchers from UCLA and Stanford call sighing "a life-sustaining reflex that prevents air sacs located in the lungs, called alveoli, from collapsing."

Study coauthor Jack Feldman told *LiveScience*, "A human lung has as much surface area as a tennis court, and so that's all folded inside your chest." If humans didn't sigh about every five minutes, the alveoli would not be able to reinflate, causing the lungs to fail. The only way to pop alveoli open again is to take a deep breath.

Now remember that sighing is just one of thousands if not millions of vital processes going on inside your body all the time. If your body was lacking even one of those vital processes, you'd be in big trouble. So the next time you sigh, thank your all-wise Creator!

Prayer: Heavenly Father, in light of what You have created, it doesn't seem possible for people to deny Your existence … and yet they do! Help them to see the foolishness of their thinking. In Jesus' Name. Amen.

Ref: L. Dodgeon, "Aaaaaaah, Really? You Would Die If You Didn't Sigh," *LiveScience*, 3/7/16.

Evolution – A Matter of Faith

John 1:3-4
"All things were made by him; and without him was not any thing made that was made. In him was life; and the life was the light of men."

Does the name Harold Clayton Urey ring a bell? Urey was an American physical chemist who did pioneering work on isotopes, earning him the Nobel Prize in Chemistry for discovering deuterium. He also played a significant role in the development of the atom bomb. Still no bell? Well, perhaps you'll recognize his name if I mention the Miller-Urey experiment, conducted in 1952.

In that famous experiment, a mixture of ammonia, methane, hydrogen and water was exposed to electric sparks to simulate lightning. The mixture ended up producing some amino acids – the building blocks of life – but many news outlets mistakenly reported that Harold Urey and Stanley Miller had created life in the lab.

Though Urey's name is still known and revered by evolutionists, I wonder how many of them know where Harold Urey ended up in his views on the origin of life. Listen carefully to what he told the *Christian Science Monitor* ten years after the Miller-Urey experiment:

"All of us who study the origin of life find that the more we look into it, the more we feel that it is *too complex to have evolved anywhere*." He went on to say he "believes as an *article of faith* that life evolved from dead matter on this planet." Did you get that? An article of faith, *not* science!

And that is exactly what evolution really is – a faith that denies both the Bible and the most fundamental laws of science.

Prayer: Heavenly Father, I pray that more evolutionists would come to realize that living cells are so incredibly complex, only our Creator could have made them! In Jesus' Name. Amen.

Ref: Quoted in Roger G. Gallop, Ph.D., *Evolution: The Greatest Deception in Modern History*, p. 46 (Red Butte Press, 2nd edition, 2014). *Christian Science Monitor*, 1/4/62, p. 4.

Biggest Volcanic Eruption Ever!

Isaiah 2:21
"To go into the clefts of the rocks, and into the tops of the ragged rocks, for fear of the LORD, and for the glory of his majesty, when he ariseth to shake terribly the earth."

Which part of the world do you suppose is home to the biggest volcanic eruption in history? Indonesia perhaps? In 1883, the volcano Krakatoa erupted, destroying over two-thirds of the island and unleashing a huge tsunami that killed more than 36,000 people. The explosion is thought to have been the loudest sound ever heard in modern history. It could be heard 3,000 miles away. That's 600 miles more than the distance between New York City and Los Angeles!

Or might it have been the Philippines? After all, over a century after Krakatoa, the eruption of Mount Pinatubo in 1991 ejected a cubic mile of material, creating a column of ash that rose twenty-three miles into the atmosphere. A year after that eruption spewed millions of tons of sulfur dioxide and other particles into the air that blocked out sunlight, global temperatures dropped by one degree Fahrenheit.

Indonesia? The Philippines? Neither! The three biggest eruptions in history all took place at Yellowstone National Park in the United States. What's more, scientists told the journal *Science* in 2013 that they discovered a huge mass of magma beneath the caldera, or "supervolcano", in Yellowstone that could fill the Grand Canyon eleven times over!

Needless to say, you don't want to be visiting Yellowstone the next time it erupts! But don't call off your vacation plans just yet. Seismologist Robert Smith from the University of Utah told *LiveScience* that the chance of an eruption at Yellowstone is only about one in 700,000 each year.

Prayer: Heavenly Father, though volcanic eruptions and other catastrophic events can cause me to fear, I know that You are in ultimate control, so I will not let my heart be troubled. In Jesus' Name. Amen.

Ref: "The 11 Biggest Volcanic Eruptions in History," *LiveScience*, 2/23/16.

Theory, Hypothesis ... Or Something Else?

2 Peter 1:16
"For we have not followed cunningly devised fables, when we made known unto you the power and coming of our Lord Jesus Christ, but were eyewitnesses of his majesty."

Does the theory of evolution pass the test of being a legitimate theory? Not according to Ph.D. geologist and marine scientist Roger Gallop. In his book *Evolution: The Greatest Deception in Modern History,* Gallop points out that a theory is "an explanation of a set of related observations based on hypotheses and verified by independent researchers." But, he adds, "evolution – that is, genuine gain in genetic information or net increase in complexity – has never been observed in fossils or living populations."

So if evolution isn't a theory, what is it? A hypothesis? Nope. Dr. Gallop writes, "Evolution is not even worthy of the term hypothesis, which is an educated guess based upon observation."

In that case, what term is appropriate? Since evolution has never been observed or substantiated by empirical science, it would be best to call evolution an assumption. Most scientists accept evolution today because they simply *assume* it to be true.

And that's not just our position. Listen to what L. Harrison Mathews wrote in his *Introduction to Darwin's The Origin of Species:* "[Since] biology is in the peculiar position of being a science founded on an unproved theory, is it then a science or a faith?" Mathews then answers the question: "Belief in the theory of evolution is thus exactly parallel to belief in special creation. Both are concepts which believers know to be true but neither, up to the present, has been capable of proof."

> **Prayer: Heavenly Father, thank You for reminding me that the theory of evolution is not a theory or a hypothesis ... and it most certainly is not a fact! In Jesus' Name. Amen.**

Ref: Roger G. Gallop, Ph.D., *Evolution: The Greatest Deception in Modern History,* pp. 4-5 (Red Butte Press, 2nd edition, 2014).

How Accurate Is Radiometric Dating?

Psalm 119:160-161
"Thy word is true from the beginning: and every one of thy righteous judgments endureth for ever. Princes have persecuted me without a cause: but my heart standeth in awe of thy word."

Evolutionists are good at giving the appearance they know what they are talking about when they say that such and such fossil is 150 million years old or 320 million years old or whatever.

Creationists have learned long ago that radiometric dating methods cannot be trusted, based as they are on faulty assumptions and incomplete information about the samples. No, we are not being anti-science. Think about it. When different labs test the same fossil sample, why do the labs often come back with results that are millions of years apart?

Let me close today's program with a story sent to us by a listener that illustrates why the old-age dates from evolutionists cannot be trusted. He writes:

"In 1973, I was stationed in Hawaii at Kaneohe Bay Marine Corps Air Station on Oahu. I found some amber-like stuff oozing from a palm tree. There were no insects in it, and it was of the consistency of chewed gum. I put a bit, the size of a marble, into a film canister. Many years later, I came upon that film canister and found the resin to be harder than wood. On a fling, a friend of ours took it to the university where he works and had it tested. The report came back that it was spruce amber ... and over 150,000 years old. I never did tell them 'the rest of the story.'"

Prayer: Heavenly Father, help me to show others why the Bible can always be trusted over the speculations of evolutionists. In Jesus' Name. Amen.

Ref: Letter on file at Creation Moments, 9/6/12.

And a Little Child Shall Lead Him

Isaiah 11:6
"The wolf also shall dwell with the lamb, and the leopard shall lie down with the kid; and the calf and the young lion and the fatling together; and a little child shall lead them."

In the early 1950s, people were stunned when former Soviet spy Whittaker Chambers testified that Alger Hiss, an American government official, was spying on the U.S. for Russia. In his book *Witness*, Chambers relates how a little child led him to change his beliefs and the course of his life.

"It was shortly before we moved to Alger Hiss's apartment in Washington. My daughter was in her high chair. I was watching her eat. She was the most miraculous thing that had ever happened in my life. I liked to watch her even when she smeared porridge on her face or dropped it meditatively on the floor. My eye came to rest on the delicate convolutions of her ear – those intricate, perfect ears."

Chambers continued: "The thought passed through my mind: 'No, those ears were not created by any chance coming together of atoms in nature (the Communist view). They could have been created only by immense design.'

"The thought was involuntary and unwanted. I crowded it out of my mind. But I never wholly forgot it or the occasion. I had to crowd it out of my mind. If I had completed it, I should have had to say: Design presupposes God. I did not then know that, at that moment, the finger of God was first laid upon my forehead."

Chambers was right. Not only does creation point to design, it points to the Designer – Jesus Christ.

Prayer: Heavenly Father, thank You for the witness of Your creation! Your grace is beyond comprehension! In Jesus' Name. Amen.

Ref: From the Foreword to Whittaker Chambers' *Witness*, first published in 1952.

What Did the First Living Cell Eat?

Colossians 1:16
"For by him were all things created, that are in heaven, and that are in earth, visible and invisible, whether they be thrones, or dominions, or principalities, or powers: all things were created by him, and for him:"

Not long ago, I was talking with another creationist about the impossibility of the first living cell coming into being through natural causes from non-living chemicals. I asked him, "Even if such a thing were possible, what would the first living cell eat?" Without missing a beat, my friend Jimmie said: "CELLery?"

The two of us shared a good laugh together, but I can assure you that the origin of life is no laughing matter to evolutionists. Many evolutionists, in fact, prefer to sidestep the issue completely, asserting that evolution deals only with the origin of different species after that first living cell has somehow come into being.

But have you ever stopped to think about how that first living cell could have survived for even a millisecond? Unless its environment was perfectly suited to sustain that cell's life, it would have died in an instant. Unless the cell had necessary nutrients, it would likewise quickly perish. And unless it had the means to reproduce itself, that first living cell would be the last. Or, perhaps, hundreds of millions of years later, the second living cell would ooze out of a primordial soup of non-living chemicals ... and once again die a millisecond later!

No, as the great scientist Louis Pasteur demonstrated in 1859, life comes only from life. Pasteur confirmed what the Bible tells us – that life on Earth came from the *living* God.

Prayer: Heavenly Father, thank You for continuing to frustrate scientists who believe that life could have come about through natural causes. Reveal to them that life was created by Your Son. In Jesus' Name. Amen.

Bacteria That Move Like "Spider-Man"

Psalm 40:5a
"Many, O LORD my God, are thy wonderful works which thou hast done...."

On a previous Creation Moments broadcast, we told you about the microscopic flagellar motor that some bacteria use to move from place to place – much like an outboard motor propels a speedboat along the surface of a lake. Today, I'm going to tell you about a different kind of nanomachine that some bacteria use to get around.

Using a new imaging technique called electron cryo-tomography, Caltech researchers were able to study the type IVa pilus machine. Professor Grant Jensen wrote that this nanomachine "allows a bacterium to move through its environment in much the same way that Spider-Man travels between skyscrapers." As Dr. Jensen explains it, the bacterium assembles a long fiber that attaches to a surface like a grappling hook. When the bacterium retracts the fiber, the bacterium literally pulls itself forward.

The professor recognizes that such nanomachines are "huge complexes comprising many copies of a dozen or more unique proteins that carry out sophisticated functions." And yet, he is unwilling to admit that such a complex machine must have come from the mind of an extremely intelligent designer. Instead, the Caltech team assumes that "over millennia, these adaptable little organisms have evolved a variety of specialized mechanisms to move themselves through their particular environments."

Unlike evolutionists who assume that sophisticated machines just sort of evolved, creationists know it was the Creator who gave even lowly bacteria such incredibly complex machines to help them move around!

Prayer: Oh Lord, I praise You for creating a universe that never ceases to fill me with awe and wonder! Amen.

Ref: California Institute of Technology, "An up-close view of bacterial 'motors'," *ScienceDaily*, 3/29/16.

The March of the Robot Mailmen

1 Chronicles 16:24-25a
"Declare his glory among the heathen; his marvellous works among all nations. For great is the LORD, and greatly to be praised..."

Imagine, if you will, a robot mailman carrying a huge sack of mail on its back, walking step-after-step on the sidewalk until it reaches its destination. You'd be pretty impressed with the robot's designer, right?

Well, such robot mailmen *do* exist, and you have trillions of them inside your body at this very moment. These robot mailmen are complex molecular machines called kinesin. As the Discovery Institute points out, kinesins "play a vital role in many cellular processes, not just transporting materials but also aiding cell replication." Through computer animation, these marvelous machines come to life in a video that the Discovery Institute has posted on YouTube. As the narrator tells us, "Masterpieces of micro-engineering, kinesins are miniature machines that carry cargo from one part of the cell to another, walking along self-assembling highways called micro-tubules."

Let me add that these robot mailmen are fast. Their two feet can take as many as a hundred steps per second! Kinesins are strong, too, carrying cargo many times their own size. And if their roadway is blocked by other cellular components, "multiple motor proteins may be used to carry a single piece of cargo together, providing enough force to break free."

Not in billions or trillions of years could such wonders come about by accident! Only one explanation can satisfy those who haven't been blinded by evolutionary indoctrination. Kinesins, like all molecular machines, are the invention of an extremely intelligent Creator and Designer!

Prayer: Father, how can evolutionists see such wonders inside the cell and not recognize they were designed? And yet, I know if it weren't for Your Holy Spirit, I'd be walking in darkness, too. Amen.

Ref: Casey Luskin, "The Workhorse of the Cell: Kinesin" video from the Discovery Institute.

The Magical Highway

Isaiah 40:3
"The voice of him that crieth in the wilderness, Prepare ye the way of the LORD, make straight in the desert a highway for our God."

On our previous broadcast, we told you about complex molecular machines called kinesins. As we mentioned, kinesins carry cargo from one part of the cell to another, walking along self-assembling highways called micro-tubules.

Today we're taking a closer look at those magical highways. Though physician Geoffrey Simmons is not a creationist, we wholeheartedly agree with this description in his book *What Darwin Didn't Know*: "Similar to a skyscraper with crisscrossing steel girders, the cell is supported by a skeleton of crisscrossing microfilaments and micro-tubules. Unlike a skyscraper, however, some of these structures are like movable scaffolding."

To give his readers a clearer picture of what is going on inside the cell, he writes: "Envision a building with changing stairwells and elevators to accommodate the changing business needs of the day. If the boss needs to check out the third floor, a stairwell will suddenly appear. If he needs to go to the basement, a chute will open up. Changes like these happen rather frequently within the cell, and the evidence indicates all the activity is purposeful."

To use our mailman analogy, micro-tubules are like sidewalks that assemble themselves just before the mailman gets to them. It really is a wonder to behold. Too bad that Charles Darwin never knew what really goes on inside the cell. If he had known, perhaps he might have thrown his *Origin of Species* into the trash.

> ***Prayer: Oh Lord, as scientists learn more and more about the things You have made, the more convinced I am that man is nothing more than an ant trying to understand how a computer works. Amen.***

Ref: Geoffrey Simmons, M.D., *What Darwin Didn't Know,* pp. 49-51 (Harvest House Publishers, 2004).

The Lizard That Sneezes Salt

Leviticus 11:30
"And the ferret, and the chameleon, and the lizard, and the snail, and the mole."

In the arid desert regions of the southwestern United States lives a creature with so many astonishing abilities, it speaks loud and clear about God's hand in creation!

Chuckwallas are large lizards that eat desert plants loaded with salt – a diet that would kill most animals. But these lizards are equipped with special glands in their nostrils that remove salt from their bloodstream. When the glands have filled up with salt, the chuckwalla sneezes, shooting crystallized salt out of its nostrils!

Our Creator has also given them the ability to change the color of their skin. Of course, other lizards and sea creatures are able to do this to hide from predators. But chuckwallas change color for a different reason. To survive in desert regions that are freezing cold at night and blazing hot during the day, their skin is dark in color in the morning to quickly absorb the warming rays of the sun. Later in the day, their skin becomes much lighter to reflect the desert heat. How cool is that?

The chuckwalla's skin is also very loose and baggy. Here's why. When a predator approaches, the chuckwalla squeezes itself into a rocky crevice. It then swallows lots of air, blowing itself up like a balloon that's wedged so tightly in the rock, the predator can't pull it out.

Now, I ask you – how can mindless, purposeless evolution account for a creature with so many extraordinary abilities?

Prayer: Heavenly Father, it's so easy to see Your Son's hand in the things He has created that it's hard to believe how anyone can be blind to it! I pray that their eyes will be opened by Your Holy Spirit! Amen.

Ref: DeWitt Steele, *Science: Order and Reality,* p. 138 (Beka Book Publications, 1980). Jobe Martin, *The Evolution of a Creationist,* pp. 249-250 (MMII Biblical Discipleship Publishers, 2004).

The Real Goal of Evolution

Romans 1:20
"For the invisible things of him from the creation of the world are clearly seen, being understood by the things that are made, even his eternal power and Godhead; so that they are without excuse:"

In their book *A Closer Look at the Evidence,* public school science teachers Richard and Tina Kleiss write that our conscience is an undeniable evidence of God's existence. Another evidence is the intricate design of nature. After all, the Bible tells us in the book of Romans that He makes Himself evident to everyone so that no one can claim that He does not exist.

And yet, many people *do* reject God's existence, basing their disbelief primarily on one thing – evolution. "If evolution is true," write the Kleisses, "there is no evidence for the existence of God from the observation of nature because the origin of everything can be explained by natural laws of science."

This is why these science educators conclude, "The real goal of evolution is removing any accountability to our Creator by denying the obvious evidence from creation that points to His existence." Yes, indeed, the evidence is so obvious, it can be seen by everyone who has not adopted the faith of evolution.

As the title of their book makes clear, Richard and Tina Kleiss invite readers to take a closer look at the evidence for God's hand in creation. This is also what Creation Moments invites you to do on each one of our radio broadcasts. When you look at the evidence all around you, God's fingerprints can be found all over the world He created.

Prayer: Heavenly Father, I pray that You will protect my loved ones from falling for the lie of evolution. Use me, Lord, to show them that You are indeed the Creator. In Jesus' Name. Amen.

Ref: Richard and Tina Kleiss, *A Closer Look at the Evidence,* March 1 (Search for the Truth Publications, 2004).

Changing Minestrone Soup into Garlic Bread

Genesis 1:11
"And God said, Let the earth bring forth grass, the herb yielding seed, and the fruit tree yielding fruit after his kind, whose seed is in itself, upon the earth: and it was so."

Creationists and proponents of Intelligent Design disagree on several matters, but we do share the same views on the utter impossibility of evolution to account for the existence of all living things. Recently we came across a wonderful explanation of how plants got here by way of evolution. It's from the book *Billions of Missing Links*, and it's so good, I just had to share it with you today.

Intelligent Design author Geoffrey Simmons writes: "The theory of evolution suggests that algae somehow arose from the primordial sea and then evolved into underwater plants, which eventually came ashore as flat plants to see more of the world or escape predators, changed to upright plants to see more of the sky, and then changed to wooded plants to see even more of the sky. Somehow they extracted carbon dioxide from a dismal ancient atmosphere that lacked oxygen and started producing oxygen for unclear reasons even though there were no animals around who used it."

Dr. Simmons writes that "this scenario through the millennia combines an incredible collection of coincidences," adding that "scores of intermediate steps – or links – are missing all along the line." He then concludes: "Changing minestrone soup into garlic bread with a change of atmosphere would be easier."

So how did plants really get here? According to the most reliable and authoritative text on this subject, God spoke them into existence on day three of Creation Week.

Prayer: Father, though scientists produce an endless number of books and articles about evolution, not one of them reveals what really happened. Thank You for revealing the truth in the Bible! In Jesus' Name. Amen.

Ref: Geoffrey Simmons, M.D., *Billions of Missing Links*, p. 113 (Harvest House Publishers, 2007).

Slightly Imperfect Sale!

2 Corinthians 5:17
"Therefore if any man be in Christ, he is a new creature: old things are passed away; behold, all things are become new."

We've all seen advertisements from retail establishments selling merchandise they call "slightly imperfect." Well, if you listen to evolutionists, we humans are more than slightly imperfect. As one writer claims, "From our knees to our eyeballs, our bodies are full of hack solutions."

The writer goes on to list such supposed design flaws as a spine that was ideal when we walked on all fours but fails miserably once we stood upright. Other so-called evolutionary flaws include: meandering arteries, crowded teeth and a retina that's backwards.

But here's our favorite "flaw" of the bunch – a misplaced voice box. "To keep food out of the trachea," he writes, "a leaf-shaped flap called the epiglottis reflexively covers the opening to the larynx whenever you swallow. But sometimes, the epiglottis isn't fast enough. If you're talking and laughing while eating, food may ... get lodged in your airway, causing you to choke." Want to hear his solution? Evolution should help us develop blowholes like whales and then move our voice box over there. Of course, he admits, we would lose our ability to talk!

While such "improved" designs from evolutionists are sometimes humorous, the Bible does tell us that we are, indeed, *seriously* flawed as a result of Adam's disobedience. Thankfully, though, God doesn't just destroy every flawed human being. He sent His flawless Son to take our many sins upon Himself to pay the penalty that we flawed and sinful people could never pay.

Prayer: Heavenly Father, I know that I have no righteousness of my own. That's why I thank You for transferring Your Son's righteousness to my account and applying my unrighteousness to His! In Jesus' Name. Amen.

Ref: Chip Rowe, "Top 10 Design Flaws in the Human Body," Nautilus, 3/24/16.

Bad Design – A Really Bad Argument!

Psalm 139:14
"I will praise thee; for I am fearfully and wonderfully made: marvellous are thy works; and that my soul knoweth right well."

On our previous program, we told you how some evolutionists enjoy finding fault with what they consider to be bad designs in nature – particularly the "poorly designed human body." They aren't using such arguments to critique evolution, of course. They are poking an accusing finger into the eye of God. If God is so smart, they say, why did He make so many mistakes?

As physician and Institute for Creation Research speaker Randy Guliuzza points out in the book *Creation Basics and Beyond*, it is evolutionists who are badly mistaken. Evolutionists, he writes, "are ignorant of the full function of the parts they criticize and are ignorant of principles governing design…. Claims that something is poorly designed are not equivalent to data-supported facts."

Dr. Guliuzza – a physician who knows the human body inside and out – also writes that evolutionists are "ignorant, from a design perspective, of the need to balance several competing interests." In fact, he adds, "design tradeoffs are actually a better indicator of intelligence behind a design."

But here's what I believe may be one of the best examples of why the bad-design argument is a bad argument and just plain silly. The late evolutionary biologist George Williams once claimed that humans having two eyes was a bad design. "Is there a good functional reason for having two eyes? Why not one or three or some other number?"

This makes me wonder if George had ever heard about depth perception.

Prayer: Heavenly Father, it's ironic that people who don't even believe You exist are so eager to criticize You! I pray that You will show them how foolish and illogical they are being. In Jesus' Name. Amen.

Ref: Randy Guliuzza, M.D., "The Mistakes in Evolutionary Arguments Against Life's Design," *Creation Basics and Beyond*, pp. 111-117 (Institute for Creation Research, 2013).

Science Turns to God's Design for Data Storage

Daniel 12:4
"But thou, O Daniel, shut up the words, and seal the book, even to the time of the end: many shall run to and fro, and knowledge shall be increased."

Do you remember when home computers used floppy disks to store data? Magnetic floppies were replaced by optical discs, using laser technology, and now we depend on hard drives, flash drives and cloud drives for the storage and retrieval of digital data. But someday even our most advanced data-storage systems will give way to the latest and greatest advances in technology.

Well, that day may soon be upon us. Researchers from Microsoft and the University of Washington have developed a system that stores digital data using synthetic DNA. They report that it can store information millions of times more compactly than current archival technologies. In fact, they claim it can "shrink the space needed to store digital data that today would fill a Walmart supercenter down to the size of a sugar cube."

According to Luis Ceze, University of Washington associate professor of computer science and engineering, "Life has produced this fantastic molecule called DNA that efficiently stores all kinds of information about your genes and how a living system works. It's very, very compact and very durable. We're essentially repurposing it to store digital data – pictures, videos, documents – in a manageable way for hundreds or thousands of years."

Once again we see scientists refusing to give credit where credit is due. DNA, the most efficient data storage system ever, was not produced by "life", whatever that may mean. DNA was designed by our life-giving Creator – Jesus Christ.

Prayer: Oh Lord, though the rapid advances in technology are impressive, nothing even begins to compare to what You designed during the six days of creation! Amen.

Ref: University of Washington, "Scientists store digital images in DNA, and retrieves them perfectly," *ScienceDaily*, 4/7/16.

Unbelievable Migrations

Mark 13:34
"For the Son of man is as a man taking a far journey, who left his house, and gave authority to his servants, and to every man his work, and commanded the porter to watch."

If you have ever traveled to a distant city on vacation, you know how much planning you have to do before you leave. And, of course, you need to know where you are going and the route you must follow to get there. But the creatures we'll be talking about today know exactly how to get where they are going. In fact, they were born with an internal GPS system to show them the way.

In his book, *Billions of Missing Links,* Dr. Geoffrey Simmons devotes an entire chapter to the topic of migration. He begins by telling how baby loggerhead turtles migrate 8,000 miles across the Atlantic Ocean. But that's a short trip compared to the staggering 25,000 miles that arctic terns fly each year. That's like flying completely around the Earth at the equator!

As Dr. Simmons points out, "Every species seems to know how to prepare for the arduous trip far in advance, but no one knows how they acquire the capability." After describing all the preparations migrating birds must take care of, he writes: "One would think all these preparations had to have come as a whole package. There is way too much purposeful change for random mutations."

How true! And even though Dr. Simmons is not a creationist, we have come to the exact same conclusion – namely, that animal migrations could not have come about slowly by trial and error, as Darwinian theory would have us believe.

Prayer: Oh Lord, though I may get lost while going on a long trip, Your creatures never seem to lose their way. You have boggled my mind once again! Amen.

Ref: Geoffrey Simmons, M.D., *Billions of Missing Links,* pp. 165-169 (Harvest House Publishers, 2007).

Why Did God Give Us Fingernails?

3 John 1:2
"Beloved, I wish above all things that thou mayest prosper and be in health, even as thy soul prospereth."

Why did God give us fingernails? So we can scratch our itches? To serve as hardhats for our sensitive fingertips? Or did our Creator give us fingernails as a diagnostic window into our bodies?

According to WebMD, the color of our nails can reveal a wide range of health issues, some of them quite serious. White fingernails, for instance, may indicate a liver disease like hepatitis. Yellowish, thickened and slow-growing nails may indicate emphysema or other lung ailment. Half-white and half-pink nails may alert doctors of kidney disease. Red nail beds may point to heart disease, while white nail beds may indicate anemia. Irregular red lines at the base of the nail may indicate lupus or connective tissue disease.

Physician Joshua Fox, spokesman for the American Academy of Dermatology, told WebMD that there is no need to run to the nearest cardiologist if your nail beds turn red. "It could very well be from nail polish," he said. The WebMD article adds, "Before assuming the worst, it's important to look at more common explanations, such as bruises, bleeding beneath the nail, and fungal infections. However, it's worthwhile to be vigilant about maintaining healthy fingernails so that you'll be alert to any potential problem."

Wasn't it good of our Creator to give us these ten tiny "windows" into the innermost workings of our bodies? God cares for us ... from the top of our heads to the tips of our fingers!

Prayer: Heavenly Father, thank You for giving me fingernails, and thank You also for physicians who can diagnose and treat diseases. In Jesus' Name. Amen.

Ref: Sherry Rauh, "Healthy Fingernails Clues About Health," WebMD, 2/24/08.

It's a Young World After All

John 17:15-17
"I pray not that thou shouldest take them out of the world, but that thou shouldest keep them from the evil. They are not of the world, even as I am not of the world. Sanctify them through thy truth: thy word is truth."

A few years ago, evolutionists discovered what they called the world's oldest mites. These little critters were encased in amber and were said to be 230 million years old. And yet, the lead researcher said, "They're dead ringers for [modern] gall mites."

Commenting on this news story, Creation Moments Board Member Dr. Don Clark said that a discovery like this does nothing to support evolution. "What it *does* support," he said, "is that God created life after its own kind *not* to change."

Dr. Clark added, "God allowed variation within a kind so there would be diversity but not life evolving from one kind to another. So it doesn't surprise the creationist that these mites show little to no change over time. That is what we would expect to see, based on God's Word."

He then observed, "If there is little to no change in the mites, perhaps the dating of the amber samples is off. Perhaps by a lot. God's Word indicates that the Earth and all of creation are 6,000 or so years old. So if these samples are only thousands of years old, would the evolutionist expect to see any change over such a short period of time? Probably not."

The friend of evolutionists is long ages ... and the more time, the better. But fossilized mites that look just like modern mites teach us that the Earth is a young, young world after all.

Prayer: Heavenly Father, even bugs encased in amber give silent testimony of a recent creation. Thank You that scientific evidences continue to pile up showing that our planet is young. In Jesus' Name. Amen.

Ref: Dr. Donald Clark interview on Broken Road Radio, 9/3/12. http://brokenroadradio.com/?p=1405

Do Birds Take a Sabbath Rest?

Genesis 2:3
"And God blessed the seventh day, and sanctified it: because that in it he had rested from all his work which God created and made."

Myles Willard is an avid bird watcher, award-winning nature photographer and long-time friend of Creation Moments. Myles has given us hundreds of breathtaking nature photos, one of which accompanies the printed transcript of today's program at the Creation Moments website.

The reason I'm telling you about him today is because of an unexpected discovery he made while looking out the window of his home in Michigan. Each fall he meticulously tracks and logs the number of migrating warblers that stop by for a rest in the big cedar tree in his yard. After tracking the activity of over 1,500 warblers for eighteen years, he was surprised to see a statistically significant dip in the number of birds stopping by *that occurred on every seventh day!*

Did these migrating birds have a built-in instinct that somehow made them follow the biblical principle of a Sabbath rest? We are not saying, of course, that the warblers were knowingly obeying God's fourth commandment. However, if God worked for six days and then rested on the seventh, why would it be hard to believe that God gave these birds a cycle of six days of work followed by a seventh day of rest?

According to the account given in the book *Inspired Evidence: Only One Reality,* "It would seem that Myles Willard, science teacher, nature photographer and bird watcher, has found and documented such a pattern."

Prayer: Oh Lord, thank You for doing all the work necessary for our salvation so we can rest securely in the knowledge that – by grace through faith – we can have eternal life! Amen.

Ref: Myles Willard, *The Rest Is History,* monograph, 2008. Cited in *Inspired Evidence: Only One Reality* by Julie Von Vett and Bruce Malone, April 29 (Search for the Truth Publications, 2012).

Which Came First – The Orchid or the Moth?

Isaiah 40:7-8
"The grass withereth, the flower fadeth: because the spirit of the LORD bloweth upon it: surely the people is grass. The grass withereth, the flower fadeth: but the word of our God shall stand for ever."

The star orchid is an amazing flower that really shouldn't even exist. You see, its pollen is located at the bottom of a long and narrow spike – or spur – that extends behind the flower. That's also where the nectar is. So how long is that spur? The last part of the orchid's scientific name gives us the answer – a whopping one and a half feet!

That these orchids even exist means there must be a pollinator that can reach into that long spur. And indeed there is. Meet the hawk moth. The proboscis – or tongue – of this moth is over ten inches long. By getting close enough to the orchid, the moth can reach the pollen with its long tongue.

When Charles Darwin received a specimen of the star orchid in 1862, he deduced that a pollinator with a tongue as long as the spur must exist. Though some people thought he was crazy, the hawk moth was discovered forty-one years later. According to evolutionists, the hawk moth "proved Darwin's theory regarding co-evolution or how plants and pollinators can influence each other's evolution."

But we would ask these people, how could the orchid survive if it arrived on the scene long before the moth? And if the moth arrived first, why would it bother to evolve such a long tongue? No, there's a more reasonable answer. They arrived at about the same time, made for each other by their all-wise Creator.

Prayer: Heavenly Father, the Bible compares us to flowers that are here for a short time and then fade away. But those who, through faith, put their trust in Your Son shall have eternal life! In Jesus' Name. Amen.

Ref: A. Grant, "Christmas Star Orchids: Tips For Growing Star Orchid Plants," Gardening Know How, 2/2/15.

Topsy-Turvy, Lefty-Righty

Matthew 7:5
"Thou hypocrite, first cast out the beam out of thine own eye; and then shalt thou see clearly to cast out the mote out of thy brother's eye."

You probably remember being taught that the lenses in your eyes turn everything you see upside down on your retina. Your brain then turns these upside-down images right side up, and all is well with the world. But wait until you hear this!

Not only are the images on your retina upside down, they are also inverted from *left to right*. As scientist and engineer Werner Gitt tells us in his book *The Wonder of Man*, "Each half of the observer's brain receives information from only one half of the image.... The left side of the brain only observes the left half of the image" – which, I might add, is really the right half of what you are looking at.

But your brain needs to do something even more incredible than sort these images out. According to Dr. Gitt, the images are also distorted "because the region around ... the fovea, where we see best ... forms an image that is ten times as large as that of the peripheral area." So your brain not only turns each half the right way up, it also needs to remove the distortion. And it performs this task beautifully! Plus, the left and right images are fused together seamlessly and without any trace of a joint!

Though evolutionists get a kick out of criticizing God for His supposedly poor design of the human eye, it is actually evolutionists who aren't seeing straight!

Prayer: *Oh Lord, how can evolutionists who say they don't even believe You exist criticize You for what they claim to be poor design of the eye? They should really be thanking You for blessing them with the gift of sight! Amen.*

Ref: Dr. Werner Gitt, *The Wonder of Man*, p. 17 (Christliche Literatur-Verbreitung e.V., 2nd English edition, 2003).

Looking for Life in All the Wrong Places

Jeremiah 29:13
"And ye shall seek me, and find me, when ye shall search for me with all your heart."

Several years ago I read a short story about three friends sitting around a campfire discussing whether life on other planets would ever be found. Two of them believe there is life on other planets, but one disagrees. He insists that God created life on their planet alone. As the conversation continues, one of them sees that the fire is dying out, so he stirs the smoldering embers with a stick he is holding in his *third green tentacle!*

Yes, I enjoy science-fiction stories! But as a creationist and Bible-believing Christian, I have my feet planted solidly on God's Word. Since the Bible does not teach that extraterrestrial life exists, neither do I.

Nevertheless, NASA recently announced they will be launching new space telescopes to look for planets capable of supporting life. According to *ScienceNews*, the first to be launched is NASA's Transiting Exoplanet Survey Satellite, or TESS. The second is the James Webb Space Telescope.

"With their powers combined," *ScienceNews* said, "TESS and James Webb could identify nearby planets that are good candidates for life. These worlds will probably be quite different from Earth – they'll be a bit larger and orbit faint, red suns – but some researchers hope that a few will offer hints of alien biology."

Creation Moments predicts they won't find what they are looking for. If scientists would only turn to searching for the *Creator* of life, they would find Him!

Prayer: Heavenly Father, I earnestly await the day when Your Son returns to the planet where He laid down His life on the cross as a sacrifice for my sins. Amen.

Ref: Christopher Crockett, "New telescopes will search for signs of life on distant planets," *ScienceNews*, 4/19/16. "The Campfire," short story by Steve Schwartz.

Meet King Kong's Opposite!

1 Samuel 17:45
"Then said David to the Philistine, Thou comest to me with a sword, and with a spear, and with a shield: but I come to thee in the name of the LORD of hosts, the God of the armies of Israel, whom thou hast defied."

Probably one of my all-time-favorite movies is *King Kong*. I'm talking about the original black-and-white film made in 1933 and starring Faye Wray. But, of course, the real star of the movie was the "eighth wonder of the world" – the enormous and powerful King Kong.

Today, however, we'll be talking about a creature that's the exact opposite of the mighty Kong in three ways. First of all, it is a *real* creature that can be found all over the western Amazon Basin as well as in Brazil, Colombia, Ecuador, Peru and Bolivia. Secondly, the creature is not an ape like the overgrown gorilla King Kong. It has a tail and is a true monkey.

And third, this creature – the pygmy marmoset – is the world's smallest monkey. Indeed, it is one of the world's smallest primates. When this creature is fully grown, its full body length – not counting its tail – is less than six inches.

Their Creator gave these tiny creatures everything they need to flourish, including special incisors to bite into trees and vines to eat the sap that comes out, and an even sharper intellect.

Like the pygmy marmoset, you might think of yourself as small and insignificant compared to others. But just as our Creator used a small boy to defeat a powerful giant with only a slingshot and a stone, so too, can God use you to accomplish great things in the power of His might!

Prayer: Heavenly Father, I know that You have used both the great and the small to accomplish great things. Lord, I pray that You will use me in whatever way You see fit. In Jesus' Name. Amen.

Ref: Wikipedia entry on "Pygmy marmoset."

Chicken or Egg – Which Came First?

Genesis 1:22
"And God blessed them, saying, Be fruitful, and multiply, and fill the waters in the seas, and let fowl multiply in the earth."

Which came first – the chicken or the egg? Well, I'm happy to report that this age-old dilemma has finally been solved. It was the chicken that came first – something that creationists have known all along. After all, the Bible tells us that God created every winged fowl and gave them the ability to reproduce after their kind.

Though atheistic evolutionists don't look for answers in the Bible, they have now come to the same conclusion as those of us who do take the Bible seriously. According to a news item in the U.K.'s *DailyMail*, "Researchers found that the formation of egg shells relies on a protein found only in a chicken's ovaries. Therefore, an egg can exist only if it has been inside a chicken."

The article adds that the OC-17 protein acts as a catalyst to speed up the development of the shell. This hard shell is essential to house the yolk and its protective fluids while the chick develops inside.

Professor John Harding, an evolutionist from Sheffield University's Department of Engineering Materials, predictably left God out of the picture when he said, "Nature has found innovative solutions that work for all kinds of problems in materials science and technology – we can learn a lot from them."

Professor Harding, the only reason why we find innovative solutions in nature for all kinds of problems is because nature itself reflects the creative genius of its Creator!

Prayer: Heavenly Father, I pray that You will remind me of today's Creation Moment whenever the conversation I'm having turns to the question of which came first – the chicken or the egg? In Jesus' Name. Amen.

Ref: "They've cracked it at last! The chicken DID come before the egg," *DailyMail*, 7/14/10.

Who Invented the Decimal Point?

Matthew 5:11
"Blessed are ye, when men shall revile you, and persecute you, and shall say all manner of evil against you falsely, for my sake."

Many evolutionists and atheists have fooled themselves into thinking that those who believe in God are incapable of making important contributions in the fields of math and science. Today we're going to take a look at just a few individuals to prove how false that claim really is.

Though you may not know who John Napier is, it was this 17th century mathematician who invented something you do know about – the decimal point. As Don DeYoung points out in his book *Pioneer Explorers of Intelligent Design,* Napier also "invented logarithms which led to the development of the slide rule." Napier didn't spend all his time dealing with numbers, however. In 1593, he published a commentary on the Bible's book of Revelation.

Then there's Leopold Kronecker, the son of Jewish parents who made important contributions to the theory of algebra, elliptical functions and calculus. After his six children embraced the Christian faith, he converted from Judaism to evangelical Christianity in the last year of his life.

Let me also mention George Boole, the British mathematician who invented Boolean Algebra. His work with binary numbers made the world of modern computers possible. Boole was also a pastor who was especially interested in creation studies.

We've run out of time today, but keep listening to Creation Moments on this station to learn about many other Bible believers who made important discoveries in such fields as mathematics, astronomy, medicine, geology and biology.

Prayer: Lord, many believers have made significant contributions in the sciences. I pray You will continue blessing the work of Your people's hands! Amen.

Ref: Don B. DeYoung, *Pioneer Explorers of Intelligent Design,* pp. 36-37, 40, 42 (BMH Books, 2006).

Are We Living in a Simulation?

2 Timothy 2:23
"But foolish and unlearned questions avoid, knowing that they do gender strifes."

If you had nothing better to do with your time, you might enjoy thinking about such things as: "How do I know that the whole world isn't just a figment of someone's imagination?" Or, "How do I know that the whole universe didn't suddenly pop into existence twenty minutes ago?"

According to an article in *Tech Insider*, "Most physicists and philosophers agree that it's impossible to prove definitively that we don't live in a simulation and that the universe is real."

No, I'm not making this up. And yes, these are questions that philosophers and evolutionists spend their time talking about. As a participant in the 2016 Isaac Asimov Memorial Debate at the American Museum of Natural History, Neil deGrasse Tyson and his fellow panelists discussed whether the universe is a simulation. If you saw the 1999 film *The Matrix*, you know what I'm talking about. Tyson said the likelihood of the universe being a simulation "may be very high," adding that "it is easy for me to imagine that everything in our lives is just the creation of some other entity for their entertainment."

What's this? Could Tyson be talking about God? As an evolutionist, this is something he would not do. Extraterrestrials maybe, but God never! While philosophers might enjoy such discussions, don't you prefer spending your time learning how the *real* universe works and giving praise to the real God who created it?

Prayer: Heavenly Father, there are so many foolish matters people waste their time on so they don't have to think about their sin and their need for salvation! Make them hungry for Your truth and restless until they find their rest in You. Amen.

Ref: Kevin Loria, "Neil deGrasse Tyson thinks there's a 'very high' chance the universe is just a simulation," *Tech Insider*, 4/21/16.

Who Invented the Periodic Table?

Ephesians 6:4
"And, ye fathers, provoke not your children to wrath: but bring them up in the nurture and admonition of the Lord."

I can still picture it in my mind – the large periodic table of the elements on the wall of my high school classroom. For those of you who don't remember it, the periodic table is a tabular arrangement of all the chemical elements, ordered by their number of protons, starting with hydrogen as number one. So who invented the table? You will find his name in the table itself. Element number 101 – Mendelevium – was named after Dmitri Mendeleev, the Russian chemist who first published the periodic table in 1869.

Back then, of course, there were far fewer known elements than the 118 elements we know about today. Interestingly enough, while he was putting the table together, he had to leave a number of gaps open for unknown elements he predicted would someday be discovered.

As Donald DeYoung points out in his book *Pioneer Explorers of Intelligent Design,* Mendeleev was raised by a devout Orthodox Christian mother who told him to "'patiently search divine and scientific truth.' ... Mendeleev thus saw chemistry as a royal and godly pursuit."

Sad to say, however, Dmitri eventually turned away from his mother's faith and embraced a form of deism. Dmitri well illustrates the truth of the old saying: "God has no grandchildren." Christian parents, though it is your responsibility to teach your children biblical truths, you must remember that their salvation is out of your hands. It is ultimately a matter between them and God.

Prayer: Oh Lord, I pray for the salvation of my loved ones. My greatest desire is that they would come to know You as their Lord and Savior! Amen.

Ref: Don B. DeYoung, *Pioneer Explorers of Intelligent Design*, p. 67 (BMH Books, 2006).

The Amazing Sea Cucumber Man!

Genesis 1:26a
"And God said, Let us make man in our image, after our likeness: and let them have dominion over the fish of the sea..."

Here's an idea for those who create superhero comic books and movies. Introducing the amazing Sea Cucumber Man – able to turn his skin into a hardened shield that can stop bullets cold!

Actually, this isn't entirely a joke. Someday scientists may be able to develop a material that would allow soldiers to wear comfortable and flexible bulletproof vests that can harden instantaneously when they are in combat situations.

The idea for such a thing comes from the sea cucumber. According to the book *Bulletproof Feathers*, the sea cucumber can crawl between narrow spaces when the whiskers of its skin are soft, but "in defensive mode, surrounding cells release molecules that cause the whiskers to bind together, forming a rigid shield."

By mimicking one of God's amazing creatures, scientists have already developed a polymeric substance that works like the sea cucumber's skin. This substance has been used to develop adaptive microelectrodes for brain implants to help doctors treat Parkinson's disease, stroke and spinal cord injuries. As the book points out, the implant is stiff when inserted into the brain and then becomes soft once it interacts with brain fluids.

Unfortunately, the book's author gives credit to nature for inspiring such cutting-edge technology and not to the God who created the natural world. That's the problem with too many scientists nowadays. They copy what God has created, but then they fail to give credit where credit is due!

Prayer: Lord, thank You for Your creation that inspires scientists to develop new technologies that benefit mankind. Though scientists seldom praise You, I will never stop giving You praise! In Jesus' Name. Amen.

Ref: Robert Allen, editor. *Bulletproof Feathers: How Science Uses Nature's Secrets to Design Cutting-Edge Technology,* pp. 42-43 (University of Chicago Press, 2010).

Tiny Flies Inspire Big Advances in Hearing Aids

Proverbs 20:12
"The hearing ear, and the seeing eye, the LORD hath made even both of them."

Do you know how your ears help you pinpoint the location of whatever you are hearing? Two factors come into play. First, you hear the sound louder in the ear that is closer to whatever is producing the sound. And second, the sound arrives at the closer ear a little sooner than the other.

So how does the tiny *Ormia ochracea* fly pinpoint sounds? Her ears are so close together, the sound arrives at one ear just a millionth of a second before the other! As Dr. Jonathan Sarfati tells us in his book *By Design*, this was no problem at all for the One who designed the fly.

God gave the tiny fly a bridge – a flexible lever – that couples both eardrums together. According to Dr. Sarfati, "The resulting resonance effectively increases the time difference about 40 times, and the eardrum nearest the sound vibrates about 10 decibels higher … which is a big difference!" Thanks to this ingenious design, "the fly's mechanical and signal-processing technology is being used to improve hearing aids, which normally can't tell direction, and it could also be used to design miniature directional microphones."

Dr. Sarfati concludes, "Once more, the design in nature has taught top human designers some lessons." And yet, scientists continue to give credit to evolution "without the slightest explanation of how the mechanical structure and nervous coding system could arise by small mutations and natural selection."

Prayer: Oh Lord, thank You for helping scientists learn from Your creation so they can, in turn, bring help to hurting people. In Jesus' Name. Amen.

Ref: Jonathan Sarfati, Ph.D., *By Design: Evidence for Nature's Intelligent Designer – the God of the Bible,* pp. 42-43 (Creation Book Publishers, 2008).

Evolutionist Defends Pedophilia

Isaiah 5:20
"Woe unto them that call evil good, and good evil; that put darkness for light, and light for darkness; that put bitter for sweet, and sweet for bitter!"

As Creation Moments has pointed out in the past, evolution is a fundamental corruption of truth with far-reaching consequences – working to undermine and subvert morality, cultures and people.

That's why we weren't surprised when outspoken atheist and evolutionist Richard Dawkins defended what he called "mild pedophilia." He says he experienced it himself as a boy, adding that other children in his school had been molested by the same teacher. He excused it, however, by saying, "I don't think he did any of us lasting harm."

The Religious News Service reports that Dawkins believes that the fear of pedophilia is overblown by society. After all, if he and his schoolmates didn't experience lasting harm from childhood sexual abuse, how bad could it be?

But, of course, Dawkins can never miss an opportunity to take a pot shot at God. He said that fundamentalist religious beliefs, like those found in the Bible, are actually a worse way to abuse a child. In Dawkins' own words, "Thank goodness, I have never personally experienced what it is like to believe…in hell. But I think it can be plausibly argued that such a deeply held belief might cause a child more long-lasting mental trauma than the temporary embarrassment of mild physical abuse."

As you can see, evolution really has far-reaching consequences that touches every area of our lives. It tosses absolute standards of morality out the door. And it can even turn an Oxford professor into a defender of pedophilia.

Prayer: Heavenly Father, while unbelievers berate You for telling us about hell, I am grateful to You for warning us about hell and for providing a Savior to keep us from going there! In Jesus' Name. Amen.

Ref: Aby Ohlheiser, "Richard Dawkins Defends 'Mild' Pedophilia, Again and Again," *The Wire*, 9/10/13.

Most Americans Are Pro-Academic Freedom

Acts 22:28
"And the chief captain answered, With a great sum obtained I this freedom. And Paul said, But I was free born."

According to a nationwide survey conducted in 2016 by the Discovery Institute, most American adults agree that "teachers and students should have the academic freedom to objectively discuss both the scientific strengths and weaknesses of the theory of evolution." In fact, 88% agree that "scientists who raise scientific criticisms of evolution should have the freedom to make their arguments without being subjected to censorship or discrimination."

Though most evolutionists in the public eye strongly oppose such academic freedom, Americans overwhelmingly agree that dissenting views in science are healthy. Just listen to these figures: 84% believe that "attempts to censor or punish scientists for holding dissenting views on issues such as evolution or climate change are not appropriate in a free society." 94% believe "it is important for policymakers and the public to hear from scientists with differing views." And 86% think that "disagreeing with the current majority view in science can be an important step in the development of new insights and discoveries in science."

According to molecular biologist Douglas Axe – who is quoted in the report – "Freedom of inquiry is a cornerstone of the scientific enterprise, but it's under increasing attack by some who claim to speak for science. So it's very good news that a large majority of Americans still believe in open and free discussion."

Unfortunately, prominent evolutionists do not agree. This makes them not only anti-academic freedom, it makes them anti-science as well.

Prayer: Father, help Your people fight for the right to oppose evolution without being silenced by those who despise freedom of thought and expression. Amen.

Ref: "Poll: Americans Overwhelmingly Support the Right of Students, Teachers, and Scientists to Discuss Dissenting Scientific Views on Evolution," Discovery Institute, 7/1/16.

Nuclear Physicist Blasts Evolution

Psalm 55:5
"Fearfulness and trembling are come upon me, and horror hath overwhelmed me."

"The denial of the Genesis account is not a matter to be taken lightly by Christians," warns nuclear physicist Brandon van der Ventel. "If the biblical record is not true, then we are left with naturalism and atheism, of which the consequences are truly horrific."

In an August 2015 interview with the Christian News Network, van der Ventel, who holds a Ph.D. from South Africa's Stellenbosch University, stressed the importance of having a biblical worldview while discussing the theory of evolution. "The final arbiter of any theory must be based on the strength of its description of physical reality. It is in this respect that the theory of evolution fails dismally."

He went on to say that physics can play a role when it comes to the sticky question of dating certain objects. "It is important to understand," he said, "that dinosaur bones, for example, are not found with a time stamp attached to them. Every publicized age is based on certain assumptions, and conflicting radioactive dates are commonplace."

According to *The Christian Post,* van der Ventel said that Christians "need to hold fast to the knowledge that God created the universe with a specific order." He argues that "if faith in the Genesis account is lost, the validity of the rest of the Bible can be questioned." He further contends that "the theory of evolution is the number-one reason why many Christians no longer believe in the creation account."

Prayer: Heavenly Father, thank You for scientists who aren't afraid to point out the weaknesses of evolution and speak out in defense of Your truth. In Jesus' Name. Amen.

Ref: Vincent Funaro, "Nuclear Physicist Insists Evolutionary Worldview 'Fails Dismally;' Says Denial of Genesis Account Should Not Be 'Taken Lightly' by Christians," *The Christian Post,* 8/4/15.

Archaeologists Find Goliath

1 Samuel 17:4
"And there went out a champion out of the camp of the Philistines, named Goliath, of Gath, whose height was six cubits and a span."

Sometimes you just have to feel sorry for Bible skeptics who cast doubt on the historical accuracy of the Bible. With each passing year, archaeologists keep finding evidences that the Bible is thoroughly reliable!

Take, for instance, the story of David and Goliath. Bible skeptics insist that the biblical account has no connection to reality. But in 2005, archaeologists, digging at the biblical home of Goliath, unearthed a shard of pottery bearing an inscription of the Philistine's name.

While the discovery does not prove Goliath's existence, it does support the Bible's depiction of life at the time of the supposed battle, said excavation director Dr. Aren Maeir of Israel's Bar-Ilan University. "What this means is that at the time there were people there named Goliath," he said. "It shows us that David and Goliath's story reflects the cultural reality of the time."

Though many skeptics tell us that David slaying the giant Goliath is a myth written down hundreds of years later, Dr. Maeir said that finding the inscription gives credence to the biblical story. As reported in Great Britain's *The Independent*, "The shard dates to about 950 BC, within 70 years of when biblical chronology says David squared off against Goliath, making it the oldest Philistine inscription found. It was at Tel es-Safi in southern Israel, thought to be the site of the Philistine city of Gath."

Clearly, the Bible's account of David and Goliath is no tall tale!

Prayer: Heavenly Father, thank You for leading archaeologists to more evidence supporting the historical accuracy of the Bible. In Jesus' Name. Amen.

Ref: Rachel Hoag, "Shard of pottery supports Bible account of David and Goliath," *The Independent*, 11/11/05.

Looking for Life on Enceladus

John 5:39
"Search the scriptures; for in them ye think ye have eternal life: and they are they which testify of me."

In an article titled "Excitement Builds for the Possibility of Life on Enceladus," a *Scientific American* writer begins with these words: "Saturn's frozen moon Enceladus is a tantalizing world – many scientists are increasingly convinced it may be the best place in our solar system to search for life."

The writer adds, "Finding life there would be a profound revelation that we are not alone in the cosmos. Furthermore, the discovery of organisms – *or the lack thereof* – could answer the subtler mystery of how life started on Earth."

In other words, whether they find life there or not, scientists are not willing to accept what the Bible tells us about how life got started on Earth. And as this article points out, scientists have now lowered the bar of what they'll accept as evidence for life. As one researcher puts it, "If you went to Mars and found a dead rabbit on the ground, it's not alive, but it is compelling evidence of life. So we're not searching for something that's alive but searching for the molecules that life uses. In other words, we're looking for the body of the dead rabbit."

One type of molecule they're looking for are amino acids. Of course, even if they do find amino acids, this isn't even close to finding evidence of life. How much wiser it would be to search for the *Creator* of life in the pages of the Bible!

Prayer: Heavenly Father, I pray that scientists involved in space exploration would give You glory for the universe that You created! In Jesus' Name. Amen.

Ref: Annie Sneed, "Excitement Builds for the Possibility of Life on Enceladus," *Scientific American* online, 6/28/16.

The Answer Is Blowin' in the Wind

Psalm 92:4
"For thou, LORD, hast made me glad through thy work: I will triumph in the works of thy hands."

When you were a child, did you ever stick your arm out of the window of your car as it traveled at high speed? If so, you will remember how the fast-moving air currents would push your arm up, down, up, down and so forth.

As an article at the Engineering 360 website tells us, "Side forces come into play whenever wind flows across an elongated object – as when an arm is projected out the side of a moving car. As the air flows around the arm, it forms vortices that come off the top and bottom of the arm in an alternating fashion. This vortex shedding, as it is called, imparts periodic forces on the arm."

The article goes on to say that this phenomenon "affects any elongated structure caught in wind or water currents, such as lampposts, high rises and the long vertical pipes used for drilling oil at sea." It can even cause a bridge to collapse – like the Tacoma Narrows suspension bridge in the state of Washington.

This is why scientists are looking for ways to reduce these forces, and they think they may have found the solution in the shape of daffodil stems. Their twisting, lemon-shaped cross-section enables daffodils to turn away from wind and protect its petals – just like their Creator planned. By studying the unique geometry of daffodil stems, engineers hope to be able to design more stable structures in the future.

Prayer: Oh Lord, thank You for Your ingenious designs that enable humans to live better and safer lives. I pray that scientists and engineers will recognize that such wonderful designs come directly from You! Amen.

Ref: "Daffodils Inspire Design of Stable Structures," Engineering 360, 5/24/16.

The Blackest Black and Whitest White

Psalm 51:7
"Purge me with hyssop, and I shall be clean: wash me, and I shall be whiter than snow."

Did you know that scientists are working to create new colors? Well, perhaps it's more accurate to say that they are working on new colorants that are being used to add color to other materials.

One such colorant is called YInMn blue – so named because it consists of the chemical elements ytrrium, indium and manganese. According to an article at the Co.Design website, YInMn blue "is the first blue pigment to be discovered since cobalt blue first entered manufacturing 200 years ago" and has the advantages of not being poisonous and not fading.

Another new pigment is called NTP yellow. Unlike YInMn blue, this stuff is very poisonous. Its technical name is lead chromate.

A third new pigment – called Vantablack – is particularly interesting because it really isn't a pigment at all. Vantablack is "essentially a specially applied coating that mimics a black hole, gobbling up light and causing it to bounce around endlessly in a forest of vertical carbon nanotubes." Created by NASA in 2011, Vantablack "was developed … to help absorb light and radiation before it could disrupt the sensors of deep space satellites."

While Vantablack is the blackest black ever, we find the whitest white in the Bible. Not only does it describe the appearance of Christ's clothing at the transfiguration, it describes those of us who have been washed in the blood of Jesus for the forgiveness of our sins.

Prayer: Heavenly Father, thank You for cleansing me from the guilt of sin through the shed blood of Your Son. In Jesus' Name. Amen.

Ref: John Brownlee, "How Scientists Invented A New Blue (And Other Colors Created In A Lab)," *Fast Company*, Co.Design, 7/19/16.

Something's Fishy About Fish Evolution

Matthew 13:47-48
"Again, the kingdom of heaven is like unto a net, that was cast into the sea, and gathered of every kind: which, when it was full, they drew to shore, and sat down, and gathered the good into vessels, but cast the bad away."

Dr. John Long, paleontologist and head of science at the Museum Victoria in Melbourne, Australia, should know a lot about fish ... and he does. But even though he wrote a book about the evolution of fish, there are many things he readily admits remain shrouded in mystery.

Dr. Carl Werner, creationist filmmaker and author of *Evolution: The Grand Experiment*, interviewed Dr. Long about fish evolution. As you're about to find out, evolutionary paleontologists are still in the dark about the origin of fish despite their total confidence that evolution is true.

Here's one quote: "The transition from spineless invertebrates to the first backboned fishes is still shrouded in mystery, and many theories abound as to how the changes took place." Here's another: "There's still a lot of debate over the origins and diversity of the first fishes. That's still one of the great mysteries and problems to be solved in vertebrate evolution." Dr. Long also said: "The mystery remains as to how sharks first evolved." And here's one more: "The origin of the bony fishes is also shrouded in mystery. So there's still mystery and some debate over where the true bony fishes came from."

Despite all these mysteries, Dr. Long is confident that many of them will be solved within the next ten years. Creation Moments predicts, however, that the mystery confronting evolutionists will only deepen. After all, fish didn't evolve; they were created by God!

Prayer: Oh Lord, I pray that You will catch many evolutionists in Your net of salvation! Amen.

Ref: Dr. Carl Werner, *Evolution: The Grand Experiment Volume 1*, p. 98 (Third Edition, New Leaf Press, 2013). Quotes are from the video *Evolution: The Grand Experiment* DVD by Dr. Carl Werner and *The Rise of Fishes: 500 Million Years of Evolution* by Dr. John Long (Johns Hopkins University Press).

Eye Shadow

Psalm 17:8-9
"Keep me as the apple of the eye, hide me under the shadow of thy wings, from the wicked that oppress me, from my deadly enemies, who compass me about."

Glass squid are amazingly good at not being seen by creatures who desire to eat them because, as their name makes clear, their bodies are transparent. In fact, as an article in *Science News* tells us, "Marine predators often scan the waters above them for the silhouettes of prey blocking sunlight, but there's little to betray a glass squid." There's just one problem, however. The squid's eyes are not transparent, exposing the creature's whereabouts.

Now, this would be a disaster for the squid if it weren't for the fact that their Creator provided some species of glass squid with silvery patches of cells called photophores. These cells are found *only* by their eyes, acting "as undersurface bioluminescence" and making "the shadows under their eyes less conspicuous to hunters below."

Evolutionists try to account for this by saying that the light-carrying photophores operate inefficiently, leaking light. This has caused one researcher to say that the cells were "really bad at being fiber-optic cables."

In other words, ask an evolutionist and they will tell you that natural selection gets the credit whether it does its job well or poorly. Sorry, but this is not science. It is a weak attempt to rationalize what can only be described as a product of ingenious design. As Creation Moments sees it, these photophores leak light for a *reason* ... and they were placed exactly where their Designer knew they would be of greatest benefit to the squid.

Prayer: Oh Lord, since You care so much about providing for the needs of even lowly creatures like the squid, I know that You will provide everything I need, both now and in eternity! Amen.

Ref: Susan Milius, "Squid stays hidden by leaking light," *Science News,* 7/9/16, pp. 12-13.

Our Amazingly Resilient Planet

2 Peter 3:13
"Nevertheless we, according to his promise, look for new heavens and a new earth, wherein dwelleth righteousness."

Many environmentalists would have us all believe that the Earth is as fragile as a raw egg. While creationists agree that we must be good stewards of the planet God has put us on, we believe He created our planet to be amazingly resilient. Even when we do something that some people feel is harmful to our planet, the natural processes that God has put into place – combined with man's own creativity – will keep our world from self-destructing.

As you know, carbon dioxide is now mistakenly considered to be a pollutant. Environmentalists are saying that the carbon dioxide being produced by mankind is putting our planet in jeopardy, so they've concentrated their efforts on reducing carbon emissions.

But now, scientists in Iceland have discovered a way to quickly turn carbon emissions quite literally into stone. According to *Science News*, researchers have found that "95 percent of the carbon dioxide injected into basaltic lava rocks mineralized into solid rock within two years." One of the study's coauthors noted, "It's working, it's feasible, and it's fast enough to be a permanent solution for storing CO_2 emissions." He also said, "We have enough basalt globally to take care of all anthropogenic CO_2 emissions, theoretically."

How ironic that volcanoes – which add so much carbon dioxide to our atmosphere with each eruption – might just turn out to provide the solution to so-called carbon pollution!

Prayer: Oh Lord, I know I can rest assured that the planet You placed me on will not be destroyed until You create new heavens and a new Earth! Amen.

Ref: Thomas Sumner, "Volcanic rock can quickly store CO_2," *Science News*, 7/9/16, p. 14.

Fish Recognize Human Faces

Genesis 42:8
"And Joseph knew his brethren, but they knew not him."

On a previous broadcast, we told you about the archerfish and its ability to shoot well-aimed squirts of water through the air at bugs. That's not all these amazing fish can do. Scientists have recently learned that they can distinguish one human face from another. Even more surprising is that the team of researchers from the University of Oxford admitted that "fish are *unlikely to have evolved the ability* to distinguish between human faces."

The article goes on to note that "human facial recognition has previously been demonstrated in birds. However, unlike fish, they are now known to possess neocortex-like structures." Now, why is this important? As researcher Dr. Cait Newport points out, "Fish have a simpler brain than humans and entirely lack the section of the brain that humans use for recognizing faces."

Dr. Newport added, facial recognition was once thought to be so difficult, it could only be accomplished by primates with large and complex brains. To see if a fish was up to the task, they tested archerfish because of their smaller and simpler brain and because these fish – and I quote – had "no evolutionary need to recognize human faces."

Yes indeed, evolution had absolutely nothing to do with a species of tropical fish being able to tell one human from another. And, we might add, molecules-to-man evolution had absolutely nothing whatsoever to do with *anything* we see in God's creation.

Prayer: Oh Lord, like the number of stars in space and the grains of sand on all the world's beaches, Your creation is filled with far too many wonders to count. I praise You for them all! Amen.

Ref: "Fish can recognize human faces, new study shows," *Phys.Org*, 6/7/16.

More Good News About Coffee

2 Timothy 3:16
"All scripture is given by inspiration of God, and is profitable for doctrine, for reproof, for correction, for instruction in righteousness:"

One of the most popular broadcasts Creation Moments has ever done was the one about the health benefits of coffee. That's not surprising when you consider all the coffee drinkers out there. However, some people did write to tell us that we should have mentioned that the caffeine in coffee is a diuretic and that coffee drinking can lead to dehydration.

But this just isn't so, according to Lawrence Armstrong, a professor at the University of Connecticut. He told *Live Science* that the idea of caffeine causing dehydration comes from a study conducted in 1928 showing that people who drank a lot of caffeinated beverages tended to go to the bathroom more often.

"The truth of the matter," he said, is that "a small increase in urine output has little to do with dehydrating the body." He added that any increase in fluid input will lead to an increase in urine output. "If you drink a liter of water, [urination] will increase," he said, but it "doesn't mean you shouldn't drink water."

Armstrong did note, however, that it is possible to consume an unhealthy amount of caffeine. But to get a lethal dose of caffeine, you would have to drink 100 cups in a day! Now, Creation Moments is not saying that caffeine is perfectly fine for everyone. However, as I'm sure many of you will agree, we can certainly thank our Creator for creating such things as the coffee bean!

Prayer: Heavenly Father, let this remind me that many things I think are true may, in fact, be incorrect. But I know for sure that every word in the Bible is true and is profitable to me in so many ways! In Jesus' Name. Amen.

Ref: "Does Caffeine Really Dehydrate You?" *Live Science*, 7/21/16.

Brain Food

Song of Solomon 4:14
"Spikenard and saffron; calamus and cinnamon, with all trees of frankincense; myrrh and aloes, with all the chief spices:"

On our previous Creation Moments broadcast, we had good news for coffee drinkers – specifically, that the caffeine in coffee does not cause a person to be dehydrated. The long-held myth that coffee is a diuretic was the product of ninety-year-old bad science. Today we've got even more good news about something you might already be consuming at breakfast. Once you hear how feeding it to laboratory mice improved their learning ability, you might want to make it a regular part of your family's breakfast.

I'm talking about cinnamon. Kalipada Pahan from Rush University Medical Center said that consuming cinnamon "would be one of the safest and the easiest approaches to convert poor learners to good learners."

He reached this conclusion after his research team discovered that feeding cinnamon to laboratory mice increased the levels of a protein called CREB in the brain's hippocampus. This is the part of the brain that is known to be the key to learning.

Pahan reported, "We have successfully used cinnamon to reverse biochemical, cellular and anatomical changes that occur in the brains of mice with poor learning." With this in mind, the research team is now investigating to see if cinnamon might be useful in the treatment of people with Parkinson's disease.

Wasn't it good of our Creator to give us so many foods and spices that are not only a joy to eat but are healthful to our bodies?

Prayer: Oh Lord, You filled the Earth with an abundance of foods and spices that are delightful and healthful. This is a gift You have given to everyone, so let everyone praise You for creating them! Amen.

Ref: "Cinnamon may aid learning ability: Spice consumption made mice better learners," *ScienceDaily*, 7/12/16. Deb Song, Rush University Medical Center press release.

Does the Fossil Record Support Evolution?

Luke 19:40
"And he answered and said unto them, I tell you that, if these should hold their peace, the stones would immediately cry out."

Ask evolutionists what they consider to be the single most important evidence for Darwinism, and you'll hear them give many different answers. But the one answer we hear most often is: "the fossil record."

Does the fossil record really support evolution? Creationists know that it actually supports biblical creation! In fact, the fossil record is a rock-solid testament in stone that billions of creatures all over the world drowned in the rising waters of the worldwide flood of Noah's time. Apparently, Charles Darwin himself realized he was unable to use the fossil record to support his theory.

On his *Evolution: The Grand Experiment* DVD, creationist Dr. Carl Werner interviewed Dr. Andrew Knoll, professor of biology at Harvard University. Here's what Dr. Knoll said about Darwin and the fossil record:

"Darwin devotes two chapters of *The Origin [of Species]* to the fossil record. And you might think that's because Darwin, like most of his intellectual descendants, would have seen the fossil record as the confirmation of his theory.... But, in fact, when you read *The Origin [of Species]*, it turns out that Darwin's two chapters are a carefully worded apology in which he argues that natural selection is correct despite the fact that the fossils don't particularly support it."

While Darwin's intellectual descendants think that the fossil record is their friend, in reality, it is one of the most important evidences for biblical creation!

Prayer: Father, I know that all the evidences in the world won't get people to stop believing in evolution or put their trust in Your Son for salvation. Only Your Holy Spirit can do that ... and I pray that He does! Amen.

Ref: *Evolution: The Grand Experiment* DVD, produced by Dr. Carl Werner, 2009 (New Leaf Publishing Group).

Warning: Space Travel May Be Hazardous to Your Health

1 Peter 5:8
"Be sober, be vigilant; because your adversary the devil, as a roaring lion, walketh about, seeking whom he may devour..."

From 1961 to 1972, the Apollo space program captured the world's attention with its eleven manned flights into space, its tragedies and its triumphs. Though the program is best remembered for the failed mission of Apollo 13, NASA succeeded beyond measure when it landed a man on the moon in 1969.

Now, after the passage of over four decades, it has been discovered that a high percentage of the Apollo astronauts who flew to the moon and back have died as a result of cardiovascular disease. Professor Michael Delp and his research team from Florida State University believe this was caused by their being exposed to radiation during their time in space beyond Earth's orbit.

The researchers found that 43 percent of the now-deceased Apollo astronauts who flew beyond Earth's orbit died because of cardiovascular problems. This is four to five times higher than the non-flight astronauts and those who had traveled into low-Earth orbit. To verify their findings, the researchers also exposed mice to the same type of radiation that astronauts experience in deep space and found that their arteries were damaged in the same way that would have led to cardiovascular disease in humans.

Let this be a reminder that Christians also have a dangerous enemy who cannot be seen and who would devour us if he could. But when we walk by faith in the risen Son of God, we need not fear our invisible enemy.

Prayer: Heavenly Father, whenever the troubles of this world start to overtake me, remind me that I do not need to fear because greater is He who is in me than he who is in the world. In Jesus' Name. Amen.

Ref: "Apollo astronauts experiencing higher rates of cardiovascular-related deaths," *ScienceDaily*, 7/28/16. Kathleen Haughney, Florida State University.

Habits Are Real Brain-Changers

Romans 8:29
"For whom he did foreknow, he also did predestinate to be conformed to the image of his Son, that he might be the firstborn among many brethren."

Ever wonder why habits are so hard to break? If you think it's just a matter of weak willpower or cravings that refuse to subside, scientists from Duke University have startling news for you. They have found that habits leave a lasting mark on specific circuits in the brain, priming you to keep engaging in the habitual activities.

While studying the behaviors of mice, including some who had developed a craving for sweets, the scientists found pathways in their basal ganglia which carry either a "go" or a "stop" signal. In mice who had developed a sweet tooth, both the "go" and the "stop" signals were more active than in the non-habit mice. In the non-habit mice, however, the "stop" signal actually preceded the "go" signal.

According to the researchers, these changes were so long-lasting and obvious, it "was possible for the group to predict which mice had formed a habit just by looking at isolated pieces of their brains in a petri dish."

Dr. Nicole Calakos, the study's senior investigator, said, "One day, we may be able to target these circuits in people to help promote habits that we want and kick out those that we don't want."

One day in the future? Christians don't have to wait. Through the power of Christ who dwells within us, we can break old habits and establish new ones that make us more like Him.

Prayer: Heavenly Father, I pray that You will break the power of habits that are not pleasing in Your sight and replace them with behaviors that conform me to the image of Christ. In Jesus' Name. Amen.

Ref: "Why are habits so hard to break?" *ScienceDaily*, 1/21/16. Duke University. Justin K. O'Hare, Kristen K. Ade, Tatyana Sukharnikova, Stephen D. Van Hooser, Mark L. Palmeri, Henry H. Yin, Nicole Calakos. "Pathway-Specific Striatal Substrates for Habitual Behavior." *Neuron*, 2016; DOI: 10.1016/j.neuron.2015.12.032.

New Antibiotic Discovered Inside Your Nose!

3 John 1:2
"Beloved, I wish above all things that thou mayest prosper and be in health, even as thy soul prospereth."

According to researchers from the University of Tübingen and the German Center for Infection Research, a bacterium in your nose may have already saved your life numerous times in the past and will likely save your life many times more in the future. That bacterium – *Staphylococcus lugdunensis* – is well known for inhabiting your nose. But now it has come to light that the bacterium produces a previously unknown antibiotic named Lugdunin.

The antibiotic is said to be "able to combat multi-resistant pathogens, where many classic antibiotics have become ineffective." According to the research team, infections caused by *Staphylococcus aureus* ranks as one of the leading causes of death worldwide. However, the researchers found that this deadly strain of Staph is rarely found when *Staphylococcus lugdunensis* is present in the nose.

According to Professor Andreas Peschel, "Normally, antibiotics are formed only by soil bacteria and fungi. The notion that human microflora may also be a source of antimicrobial agents is a new discovery." The discovery is good news in itself. Even better, it may open up new ways to develop strategies for infection protection and to find new antibiotics within the human body.

Who could have thought that microscopic bacteria in our noses produce an antibiotic that could protect us from deadly microbes? There's no question in my mind that our Creator thought of it because He cares so much about us!

Prayer: Heavenly Father, I praise You for providing me with life, breath and protection from deadly diseases. I praise You most of all for sending Your Son to Earth to die in my place for the forgiveness of my sin. Amen.

Ref: University of Tübingen. "Human nose holds novel antibiotic effective against multiresistant pathogens." *ScienceDaily*, 7/29/16.

The Father of Entomology

Isaiah 7:18
"And it shall come to pass in that day, that the LORD shall hiss for the fly that is in the uttermost part of the rivers of Egypt, and for the bee that is in the land of Assyria."

More than 150 years after his death in 1850, William Kirby is still known as one of the first scientists to devote his life to the study of insects. Let me add that he was also devoted to something else. Shortly before publishing his first major work on the bees of England, Kirby wrote the following:

"The author of Scripture is also the author of Nature: and this visible world, by types indeed, and by symbols, declares the same truths as the Bible does by words. To make the naturalist a religious man – to turn his attention to the glory of God, that he may declare his works, and in the study of his creatures may see the lovingkindness of the Lord – may this in some measure be the fruit of my work…"

Between 1815 and 1826, Kirby and fellow British entomologist William Spence coauthored the four-volume *An Introduction to Entomology: or Elements of the Natural History of Insects*. These books are still regarded as the foundational work in the field of entomology. In 1835, Kirby wrote *The History, Habits and Instincts of Animals*. In the first chapter – titled "Creation of Animals" – he writes that the existence of animals testifies to the existence of their Creator.

If William Kirby, the father of entomology, had lived nine years longer to see the publication of Charles Darwin's *Origin of Species*, I'm quite sure he would have been an outspoken opponent.

Prayer: Oh Lord, though evolutionists want us to believe that all true scientists are evolutionists, I thank You for the many scientists – both living and dead – who recognize that You are the Creator! Amen.

Ref: https://en.wikipedia.org/wiki/William_Kirby_(entomologist).

Did Dinosaurs Stop to Smell the Roses?

Isaiah 40:8
"The grass withereth, the flower fadeth: but the word of our God shall stand for ever."

    For years, evolutionists have told the public that during the time of the dinosaurs, plants were only of the non-flowering kind. According to these experts – including the most famous evolutionist of his time, Carl Sagan – flowering plants had not yet evolved. Dinosaurs only saw non-flowering cone trees and cycads.

    When Dr. Carl Werner heard Carl Sagan say that "dinosaurs perished around the time of the first flower" on his *Cosmos* television series, he reasoned that if plants are, indeed, evolving, then evolution must be a fact.

    To verify the accuracy of what he and 500 million television watchers had been told, Dr. Werner began his own investigation, eventually taking him and his wife to museums and dig sites all over the world. In his travels, he discovered abundant evidence of perfectly preserved flowering plants in the same rocks where dinosaurs are found. As he describes it in his book and DVD *Living Fossils,* his first reaction was: "What? Rhododendrons living with dinosaurs? Debbie and I have these growing in our backyard!"

    Over time, he found poppies, lily pads, leaves from sassafras, sweetgum and poplar trees as well as many other flowering plants in *Cretaceous* rocks that looked exactly like modern-day plants. His investigation shows that dinosaurs walked among the rhododendrons and flowering poppies just as the Bible's account of creation suggests. For all we know, the fearsome *T. Rex* might have even stopped to smell the roses!

Prayer: Heavenly Father, I pray You will make it clear to Your people that much of what they have been taught by evolutionists is simply not supported by the evidence and is contrary to what the Bible teaches. Amen.

Ref: *Living Fossils*, Vol. 2 of *Evolution: The Grand Experiment*, pp. 210ff (2008, New Leaf Press). *Living Fossils*, Vol. 2 of *Evolution: The Grand Experiment* DVD (2011, New Leaf Press).

What's the Smallest Thing You Can See?

> **Psalm 94:9**
> *"He that planted the ear, shall he not hear? he that formed the eye, shall he not see?"*

Evolutionists never get tired of mocking the supposedly poor design of the human eye. They are attempting to poke a finger into the eye of the Christian's Creator God when they say, "If God is such a good designer, why did He do such a sloppy job with the eye?"

On previous broadcasts, we have provided many reasons why the human eye is one of God's most magnificent creations. Today we'll take a look at new research showing that the human eye can detect the tiniest speck of light there is – a single photon.

Previous experiments have shown that humans can see blips of light made up of just a few photons. But according to new research published in *Nature Communications,* the subjects were able to detect a single photon – something the researchers never expected.

Physicist Alipasha Vaziri of Rockefeller University in New York City says, "If you are somewhere outside of a city in nature and on a moonless night and you have only stars to navigate, on average the number of photons that get into your eye is approaching the single photon regime." This is why he feels that having eyes sensitive enough to see single photons "may have some evolutionary advantage."

It's not surprising that evolution gets all the credit. However, whether they think the eye was poorly or beautifully designed, evolutionists must recognize that the eye was *designed* ... and this implies a Designer!

> **Prayer: Oh Lord, thank You for opening my eyes to see so many wondrous things in the Scriptures and in the universe You created! Amen.**

Ref: Emily Conover, "Human eye spots single photons," *ScienceNews*, 7/28/16. J.N. Tinsley et al. "Direct detection of a single photon by humans," *Nature Communications*. Vol. 7, 7/19/16. doi: 10.1038/ncomms12172.

Fisherman's Dream

Jonah 1:17
"Now the LORD had prepared a great fish to swallow up Jonah. And Jonah was in the belly of the fish three days and three nights."

For me, catching a fish of any size is a great accomplishment, but can you imagine what it would be like to find a gigantic Mola mola at the end of your fishing line? Also known as the ocean sunray, the Mola mola is the world's largest bony fish, weighing as much as 5,000 pounds and measuring up to fourteen feet from top to bottom!

Scientists aboard Exploration Vessel Nautilus recently spotted and photographed one of these gigantic fish in the eastern Pacific ocean. One researcher told *Live Science*, "They fascinate me because there is so little known about them despite their large size."

Another researcher agrees. Rich Bell, a fisheries scientist aboard the research vessel, said: "Their actual biology is relatively little known. Mating, their growth, migration patterns, are not particularly well-known."

And yet, *Live Science* wasted no time in saying that the fish – which scientists know so little about – "is considered to be evolutionarily advanced as they are thought to be one of the most recent fish families."

Now, as I'm sure you recognize, this isn't a scientific statement at all. It is a statement of faith. By their faith in Darwinism, they simply assume that evolution gave rise to this enormous fish. Creationists, too, exercise faith, but it is faith in the Author of the never-failing Word of God. And that's why we say with full assurance that the Mola mola comes from the hand of our Creator.

Prayer: Heavenly Father, I know that all living things have their origin in Your Son. From land creatures to fish that live in the sea, He created them all! In Jesus' Name. Amen.

Ref: Kacey Deamer, "Holy Mola: Scientists Spot World's Largest Bony Fish," *Live Science*, 8/1/16.

Waters Shall Be Turned to Blood

Exodus 7:17
"Thus saith the LORD, In this thou shalt know that I am the LORD: behold, I will smite with the rod that is in mine hand upon the waters which are in the river, and they shall be turned to blood."

If you're listening to today's broadcast on the radio, let me suggest that you go to the Creation Moments website and do a search on the title of today's program: "Waters Shall Be Turned to Blood." You will see an incredible satellite photograph showing Lake Urnia in Iran looking like it's filled with blood rather than the green waters that were there just a few months earlier.

As reported in *Live Science*, the color change is likely caused by microorganisms in the lake that thrive on salt and light. According to the NASA Earth Observatory, scientists point to a bacteria family called *Halobacteriaceae* and the algae family *Dunaliella* as the most likely suspects for Lake Urnia's current crimson color.

Live Science points out that many other bodies of water have turned red from time to time. Now, I know what you may be wondering. Could it have been bacteria that caused the River Nile to turn blood-red at the exact moment that Moses struck the waters with his rod? Could bacteria have been the means by which God accomplished the miracle? Frankly, we don't believe so because the Bible says specifically that the waters turned to blood. If the Nile had only become blood-red in appearance, Moses could easily have written it that way.

But this we know with certainty – that Jesus shed real blood on the cross for the salvation of all those who put their trust in Him.

Prayer: Oh Lord, thank You for shedding Your blood so that my sins – past, present and future – can be forgiven. Amen.

Ref: Mindy Weisberger, "'Blood Lake' Blooms in Iran," *Live Science*, 7/27/16.

Just a Minute

Hebrews 9:28
"So Christ was once offered to bear the sins of many; and unto them that look for him shall he appear the second time without sin unto salvation."

According to the *Journal of the American Heart Association*, it takes just one minute of breathing secondhand marijuana smoke to cause blood vessel impairment for the following ninety minutes. That is, if you're a rat. But Dr. Matthew Springer, senior author of the study, noted, "Arteries of rats and humans are similar in how they respond to secondhand tobacco smoke, so the response of rat arteries to secondhand marijuana smoke is likely to reflect how human arteries might respond."

The research team noted that one minute of breathing secondhand tobacco smoke impairs the arteries for the next thirty minutes, making marijuana smoke three times more damaging to those who breathe it.

Dr. Springer added, "While the effect is temporary for both cigarette and marijuana smoke, these temporary problems can turn into long-term problems if exposures occur often enough and may increase the chances of developing hardened and clogged arteries."

The same can't be said of sin, however. Even just one minute of sin can lead to a lifetime of regret. In fact, mulling over the very thought of disobeying God is tantamount to the sinful act itself. Surely, all of us need a Savior to save us from the relentless power and penalty of sin. This is why Jesus came to die on the cross. If you haven't already asked Jesus to be your sin-bearer, don't put it off a minute longer!

Prayer: Heavenly Father, I know that salvation is something I could never earn, so thank You for sending Your Son to die on my behalf! In Jesus' Name. Amen.

Ref: "A minute of secondhand marijuana smoke may damage blood vessels: Study in rats," *ScienceDaily*, 7/27/16. Source: American Heart Association.

New Glue from Ivy

Romans 8:39
"Nor height, nor depth, nor any other creature, shall be able to separate us from the love of God, which is in Christ Jesus our Lord."

All of us have seen ivy clinging to the sides of houses and trees, but did you know that these climbing plants can withstand the force of hurricane and tornado-force winds? What gives them such sticking power? Well, an article at the Engineering 360 website points out that scientists are now learning that ivy "might hold the key to manufacturing stronger adhesives, more durable paints and even cosmetics with better staying power."

According to Mingjun Zhang, biomedical engineering professor at Ohio State University, "When climbing, ivy secretes … tiny nanoparticles, which make initial surface contact. Due to their high uniformity and low viscosity, they can attach to large areas on various surfaces." Then after the moisture evaporates, a chemical bond forms.

Zhang and his research team have used the nanoparticles to reconstruct a simple glue that mimics ivy adhesive. Ultimately, he is interested in creating bio-adhesives that can aid in repairing wounds.

Since ivy is also a pest plant that can be destructive to buildings and bridges, Zhang hopes that his team will come up with an approach to keep ivy from attaching itself to structures.

Though most types of glue lose their stickiness over time, there is one kind of "glue" that is guaranteed to work forever. We find it in the Bible, where it says that nothing can separate us from the love of God, which is in Christ Jesus, our Lord.

Prayer: Heavenly Father, thank You for loving me with a love that will endure forever. In Jesus' Name. Amen.

Ref: John Simpson, "Chemistry Behind Ivy's Powerful Grasp Could Yield Stronger Adhesives," Engineering 360, 6/16/16.

Second Skin for Soldiers

Proverbs 29:25
"The fear of man bringeth a snare: but whoso putteth his trust in the LORD shall be safe."

While Creation Moments is outspoken in criticizing evolution, we are just as vocal about extolling the work of scientists who are doing *real* science – especially the science that benefits mankind. For example, scientists are now working on developing a "second skin" for soldiers that would protect them from biological and chemical weapons.

Lawrence Livermore National Laboratory has already developed a material that they say is "highly breathable yet protective from biological agents" and that it's "the first key component of futuristic smart uniforms that also will respond to and protect from environmental chemical hazards."

This new material is made of polymeric membranes with aligned carbon nanotube channels. The size of these channels, or pores, is less than 5 nanometers, which is 5,000 times smaller than the width of a human hair. The pores are also half the size of biological threats like bacteria and viruses, so it would keep them out ... but they are large enough to allow air and water vapor to pass right through.

Group leader, Kuang Jen Wu, said the fabric will be able to block mustard gas and "toxins such as staphylococcal enterotoxin and biological spores such as anthrax." The research team expects that the new uniforms could be deployed in the field within ten years.

Christians, however, don't need to wait that long to be protected. That's because those who put their trust in Jesus are already safe in His care.

Prayer: Heavenly Father, I know that I don't need a second skin. I need a second birth. Thank You for sending Your Son to die in my place so I could be born again. In Jesus' Name. Amen.

Ref: Lawrence Livermore National Laboratory. "'Second skin' protects soldiers from biological, chemical agents," *ScienceDaily*, 8/3/16.
<www.sciencedaily.com/releases/2016/08/160803095341.htm>.

Parrot to Testify at Murder Trial?

Numbers 35:30
"Whoso killeth any person, the murderer shall be put to death by the mouth of witnesses: but one witness shall not testify against any person to cause him to die."

Should a parrot be allowed to testify at a murder trial if the bird appears to be an *ear*-witness to the crime? That's a question that is ruffling the feathers of animal experts and legal eagles alike!

After all, African grey parrots are very intelligent. They can speak with enormous vocabularies, and they even demonstrate inferential reasoning, reports *Live Science*. The problem is – people aren't sure if Bud, the parrot, is repeating a conversation he heard during an actual murder or something he heard on television.

According to the *Detroit Free Press*, the parrot has been saying in a man's voice, "Get out!" followed by a woman's voice saying, "Where will I go?" Then, in a man's voice, "Don't – expletive deleted – shoot!" The prosecuting attorney said he wasn't aware of a precedent allowing a parrot into a trial, but he would look into whether Bud's testimony could be presented as admissible evidence.

Most experts doubt that Bud can give reliable testimony. According to Harvard psychologist Irene Pepperberg, "Basically, the issue is whether a parrot can learn a phrase that it has heard only once." She added, however, that "the only evidence that a bird could learn a phrase heard once, under stress, comes from [the deceased] Nobel Laureate Konrad Lorenz."

So what do you think? Personally, I don't think the parrot should be allowed to testify. I mean, how do you swear in a parrot?

Prayer: Oh Lord, parrots are such amazing creatures. Let me always praise You for the many wonders You have created! Amen.

Ref: Laura Geggel, "Polly Says What?! Should Parrots Testify at Murder Trials?" *Live Science*, 6/30/16.

Hurry Up, Blood Cells!

Leviticus 17:11
"For the life of the flesh is in the blood: and I have given it to you upon the altar to make an atonement for your souls: for it is the blood that maketh an atonement for the soul."

Over the years, Creation Moments has presented many interesting facts about blood, but a new study from the University of Rochester Medical Center may be the most amazing one of all.

As reported in *ScienceNews*, researchers have now discovered that when the brain runs low on oxygen, "red blood cells sense the deficit and hurl themselves through capillaries to deliver their cargo." The science journal also noted that this research on mice suggests that the cells "can both detect and remedy low oxygen."

So how exactly do red blood cells remedy an oxygen deficit in the brain? According to Jiandi Wan from New York's Rochester Institute of Technology, when oxygen levels run low, "red blood cells pick up speed by becoming more flexible." This "bendiness" allows the cells "to squeeze through narrow capillaries faster." Only when the researchers used chemicals to stiffen the red blood cells did the cells slow down.

Study coauthor Maiken Nedergaard of the University of Rochester Medical Center said that these findings might ultimately be applied to treating disorders where the link between neural activity and blood flow is damaged, including Alzheimer's disease.

This is exciting news, but it isn't good news for evolutionists. How can they explain this amazing property of red blood cells? They can't. Creationists, however, have no difficulty explaining it. But even *we* are at a loss for words when we consider that Jesus willingly shed His own blood to save the lost.

Prayer: Heavenly Father, thank You for sending Your beloved Son to pour out His blood on the cross so that we might be forgiven! In Jesus' Name. Amen.

Ref: Laura Sanders, "Red blood cells sense low oxygen in the brain," *ScienceNews*, 8/4/16.

Screws and Nuts in a Weevil's Legs

1 Corinthians 1:20
"Where is the wise? where is the scribe? where is the disputer of this world? hath not God made foolish the wisdom of this world?"

On a previous broadcast, we told you about a small hopping insect with a set of gears in its legs, allowing both legs to jump at the exact same moment. Today, I bring to your attention the first animal ever found to have a screw-and-nut mechanism in its body. I'm talking about the weevil.

A few years ago, *New Scientist* reported how a scientist made this discovery while studying CT scans of a weevil species from New Guinea. He was "surprised to find that its legs appeared to be *screwed into its body*."

That's right! The top segment of each leg is attached to a small part called the trochanter ... and the trochanter attaches the leg to the insect's body by screwing into a part called the coxa. On the inside of the coxa and on the outer surface of the trochanter, the scientist "found ridges just like those on screws and nuts."

So we see that our Creator invented screws and nuts, not mankind. But, of course, the *New Scientist* article claims, "The weevils are another example of evolution coming up with the same solutions to problems as human engineers."

Oh really?! The screws and nuts you'd buy in a hardware store were designed and made by *intelligent* beings. In the same way, the screws and nuts in the weevil's legs must have been designed and made by an intelligent being as well. Any other explanation is, well, just plain nuts!

Prayer: Oh Lord, thank You for giving me eyes to see the wonders of Your creation! I pray that You will give me boldness to share these wondrous things with everyone You put in my path. Amen.

Ref: Michael Marshall, "Beetles beat us to the screw and nut," *New Scientist*, 6/30/11.

Why Do We Smile?

Ecclesiastes 3:4
"A time to weep, and a time to laugh; a time to mourn, and a time to dance;"

Did you know it's practically impossible to laugh without smiling at the same time? Try it sometime. Oh, you can make laughing sounds without smiling, but you can't laugh for real. Well, if you're anything like me, you just might laugh out loud when you hear what evolutionists are now telling us about why babies smile and laugh.

According to evolutionists at Kyoto University, babies smile not because they are amused or because they are trying to communicate with their parents. Babies smile for the same reason that monkeys smile – "to facilitate the development of cheek muscles, enabling humans, chimpanzees, and Japanese monkeys to produce smiles, laughs, and grimaces." They add, "Spontaneous smiles don't express feelings of pleasure in chimpanzees and Japanese monkeys; rather, the smiles are more similar to submissive signals and grimaces rather than smiles."

According to study author Masaki Tomonaga, "We can infer that the origin of smiles goes back at least 30 million years, when old world monkeys and our direct ancestors diverged."

Really now! Do evolutionists honestly expect us to believe this? There's nothing funny about evolutionary nonsense like this. The reason we smile – even when we're babies – is because God created human beings as emotional creatures capable of expressing an incredibly wide range of feelings. When's the last time you thanked God for such a wonderful gift?

Prayer: Heavenly Father, I know there is a time to weep, a time to laugh, a time to mourn and a time to dance. But loving Your Son for His great sacrifice on my behalf is something I can do all the time! In Jesus' Name. Amen.

Ref: "Smiling baby monkeys and the roots of laughter," *ScienceDaily*, 8/4/16. Fumito Kawakami, Masaki Tomonaga, Juri Suzuki. The first smile: spontaneous smiles in newborn Japanese macaques (Macaca fuscata). *Primates*, 2016; DOI: 10.1007/s10329-016-0558-7.

Fast Food

Hebrews 4:16
"Let us therefore come boldly unto the throne of grace, that we may obtain mercy, and find grace to help in time of need."

Many fish are surprisingly fast when they're hunting for lunch. Which species do you think is the fastest-swimming fish in the ocean?

Some people say it's the sailfish. Sailfish have been clocked at a speed of 68 miles per hour while leaping. Others say black marlins hold the record. According to BBC Video, marlins are capable of swimming at speeds up to an incredible 80 miles per hour! But today's broadcast is about another fish that's been clocked doing 80.

I'm talking about the swordfish. Recently, scientists have discovered the secret behind the swordfish's incredible speed. For a long time, scientists have known that the swordfish's bill – or "sword" – is porous and rough. This reduces drag in the water. But now a scientist has discovered that swordfish have a gland – located where their "sword" attaches to their skull – and this gland secretes water-resistant oil which coats the swordfish's skin. According to Dr. John Videler, a biologist and professor at Groningen University in the Netherlands, the oil reduces drag in the water by 20 percent!

Until now, it was thought that the gland played a part in the fish's olfactory system. But these new findings make it more clear than ever that swordfish were designed and built for speed. Our Creator gave them powerful, streamlined bodies and everything else they need … just as He has given us everything we need!

Prayer: Oh Lord, Your creatures take my breath away! How can evolutionists fail to see Your hand in creation? Open their eyes, Jesus. Amen.

Ref: Mindy Weisberger, "Secret to Swordfish's Speedy Swimming Found," *Live Science*, 7/7/16.

The Great Pretender

Psalm 118:8-9
"It is better to trust in the LORD than to put confidence in man. It is better to trust in the LORD than to put confidence in princes."

As we have reported on previous Creation Moments broadcasts, some snakes hide when they are threatened. Others play dead, hoping predators will lose interest. Some snakes rear up or make hissing and rattling sounds to scare off would-be predators. But have you ever heard of a harmless caterpillar that can make itself look and act just like a deadly snake?

If you live in Costa Rica, Belize, Mexico or Guatemala, one of the most unusual creatures you will ever come across is the *Hemeroplanes triptolemus* moth while it's still a caterpillar. Though it looks pretty ordinary, when it is threatened by a predator, it expands and raises its body while twisting it upside-down, revealing what looks like a fearsome snake's head! Not only does the caterpillar look like a snake, it even acts like one, harmlessly striking at predators as if it could kill. Even its shiny eye spots are menacing!

Evolutionists, of course, will tell you that these unusual markings and snake-like behavior are the result of millions of years of chance mutations. But to anyone who hasn't been fooled by such explanations, it is obvious that this caterpillar was designed.

That's why it's so important that we teach our young people to think for themselves and not blindly accept what evolutionists are telling them. Trickery and deception are common throughout the animal kingdom. But the most deceptive creature of all is man.

Prayer: Heavenly Father, I pray that You will protect me and my loved ones from those who are out to deceive us and rob us of the hope we have in Christ. Remind us to turn a deaf ear to those who deny that Your Word is truth. Amen.

Ref: Sarah Griffiths, "Caterpillar looks like a snake to scare off predators," *Daily Mail*, 5/28/14.

Letting God Create Your Day, Volume 8 Index

Page	Title
119	Always Be Ready
265	Amazing Sea Cucumber Man, The
85	Amazing Underwater Eyes
134	Ancient Astronauts
176	Ancient Birds Flew Over Dinosaurs' Heads
55	Ancient Romans and Nanotechnology
242	And a Little Child Shall Lead Him
100	Another Attempt to Erase Jesus from History
272	Answer Is Blowin' in the Wind, The
97	Anti-Theory of Evolution, The
59	Antzilla!
270	Archaeologists Find Goliath
23	Are Animals Just as Smart as Humans?
218	Are Biblical Creationists a Stumbling Block?
263	Are We Living in a Simulation?
50	Are You Superstitious?
244	Bacteria That Move Like "Spider-Man"
251	Bad Design – A Really Bad Argument!
27	Bad News, Worse News, Good News
9	Bats Cry: That's Mine!
168	Beautiful But Deadly Woman
172	Best Camouflage of All, The
239	Biggest Volcanic Eruption Ever!
233	Bigmouth!
99	Bird Dung Spider, The
231	Bird You Must Hear to Believe, The
24	Black or White?
273	Blackest Black and Whitest White, The
206	Blood – Another Masterpiece of Design
279	Brain Food
68	Brilliance of Butterfly Wings, The
142	Bubonic Plague, a Problem for Evolutionists
175	Bullies!
226	Can You Wiggle Your Ears?
145	Carbon-14 Being Found in Dinosaur Fossils!
126	Case Study in Scientific Fraud, A
225	Caution: Fuzzy Words Ahead!
249	Changing Minestrone Soup into Garlic Bread
155	Charles Darwin and Karl Marx
6	Chicken from Hell, The
261	Chicken or Egg – Which Came First?
57	Cloudy with a Chance of Crayfish
120	Coat of Many Colors
190	Coffee – Brimming with Health Benefits
69	Come Out, Come Out, Wherever You Are!
156	Comfort from Biblical Creation

Page	Title
73	Confrontation with a Radical Atheist
96	Cosmologist Paul Davies on Faith
188	Could a Magnet Separate You from God?
195	Creation Adventures Near You
15	"Creation Is a Scientific Fact"
183	Creation Truth in a Comic Strip
217	Creationist Astrophysicist
110	Darwin Doubters Must Be Punished!
12	Darwin's Dilemma Solved?
187	Darwinism Rests Its Case on a Lawyer's Claims
165	Darwin's Headhunters
79	Dawkins Becomes a Creationist
137	Death Knell of Christianity?, The
106	Designed to Stand
11	Did Comets Help Create Life?
285	Did Dinosaurs Stop to Smell the Roses?
198	Did Our Faces Evolve for Being Punched?
197	Did Our Hands Evolve for Fighting?
66	Did We Once Walk on All Fours?
46	Digging Up the Dirt on Fraudulent Fossils
133	Dinosaur Feathers!
18	Dinosaurs Are Only Thousands of Years Old
53	Dinosaur-to-Bird Evolution Challenged
49	Disposable Diapers for Robins
179	Do Americans Still Believe in a Creator?
256	Do Birds Take a Sabbath Rest?
196	Do Lemmings Commit Suicide?
103	Doctor Moses
280	Does the Fossil Record Support Evolution?
90	Don't Have a Cow, Man!
125	Don't Mess with the Hoopoe!
56	Drake Equation, The
30	Dung Beetle Navigation
17	Earthworm or Draining Bathtub?
180	Easter Island Heads – Not Just Another Pretty Face
194	Eau de Whale Guts
121	Eggs-ellent Examples of Design
182	Esteemed Creationists Defend a Young Earth
238	Evolution – A Matter of Faith
191	Evolution Illusion, The
63	Evolution in Reverse
112	"Evolution Is Effectively Dead"
41	Evolution Produces Crybabies
227	Evolutionary Hype!
267	Evolutionist Defends Pedophilia
16	Evolutionist Rebukes Evolutionists
148	Evolutionists – Masters of Mimicry
70	Evolutionists Are Simple Minded

216	Evolutionists Still Clueless on Origin of Life
102	Evolutionists Still Defend Abiogenesis
275	Eye Shadow
171	Eyewitness Account of the Exodus?
232	Eyewitness to a Volcano's Birth
13	Fascinatin' Rhythm
296	Fast Food
158	Fastest Tongue in the West
284	Father of Entomology, The
34	Feathered Dinosaurs
214	Fermi Paradox, The
21	Fireflies Light Path to Brighter Bulbs
76	First Hero in the Bible's Hall of Faith?
164	Fish Fountain, The
277	Fish Recognize Human Faces
54	Fish that Predicts Earthquakes, The
207	Fish with Three Lines of Defense, The
287	Fisherman's Dream
20	Fishy Sign Language
140	Five Second Rule Scientifically Tested, The
31	Fly with Ants on Its Wings, A
95	Fossil Jaw Not Evidence for Evolution
138	Four Eyes Or Two?
222	From Earth to Mars in Three Days
147	Giant Redheaded Centipede
150	Glow, Little Cockroach
224	God's Special Gift to Salmon
219	Goldilocks Zone, The
297	Great Pretender, The
86	Greatest Mass Murderer of All Time, The
282	Habits Are Real Brain-Changers
128	Half of Scientific Literature May Be Untrue
141	Hallucigenia
223	Has Evolution Produced Anything of Practical Value?
152	Hawking Joins Search for Extraterrestrial Life
115	Healing Power of Prayer Undeniable
241	How Accurate Is Radiocarbon Dating?
234	How Cats Really Drink Milk
210	How Do Bats Land Upside Down?
19	How Many Smells Can You Smell?
162	How Tears Point to a Designer
143	How The Seahorse Got Its Tail
230	Human Hand Is Evidence of a Creator
42	Humans with Tails?
293	Hurry Up, Blood Cells!
166	Immigration Problem
208	Implausibility of Evolution, The
160	Importance of a Worldwide Flood, The

91	In the Blink of an Eye
75	Infinite Monkey Theorem, The
204	Innocent Blood
43	"Invisible" Mouse, The
114	"Israel Owes Us for the Ten Plagues!"
255	It's a Young World After All
139	Jet-Propelled Squid
127	Jumping to Conclusions
117	Jumping Tomatoes
289	Just a Minute
84	Knocking Evolution Over with a Feather
51	Knotty Solution, A
184	"Leaves of Three, Let Them Be"
111	Life Thrives Under Antarctica's Ice
71	Light Bends!
107	Liquid Gold
36	Little Dipper, The
108	Living Corkscrew, The
5	Living Gears!
247	Lizard That Sneezes Salt, The
259	Looking for Life in All the Wrong Places
271	Looking for Life on Enceladus
246	Magical Highway, The
229	Make Noise, Not War
245	March of the Robot Mailmen, The
74	Master of DisgEYES!
32	Maxwell's Prayer
163	Meet a Creationist Who's Also a Nuclear Physicist
260	Meet King Kong's Opposite!
60	Mercury Drives Zebra Finches Crazy
199	Metal That's Light as a Feather
109	Michael Crichton on Consensus Science
278	More Good News About Coffee
186	Mortimer Adler's Anti-Darwinism Crusade
268	Most Americans Are Pro-Academic Freedom
157	Mr. Peanut
169	NASA's Glory Has Departed
283	New Antibiotic Discovered Inside Your Nose!
290	New Glue from Ivy
10	New Type of Eye Discovered!
193	Niagara Falls – Evidence for a Young Earth
221	Night of the Living Zombie Ant!
167	No Brain? No Problem!
118	No Obvious Signs of Alien Life
181	"No Real Scientist Believes in Creation"
189	"No, Thanks. I'd Rather Walk."
177	Now You See It, Now You Don't!
269	Nuclear Physicist Blasts Evolution

129	Old Shell Game, The
64	On Bended Knees
235	One of Evolution's Best-Kept Secrets
65	Open-and-Shut Case for Creation, An
213	Ota Benga: Man or Monkey
276	Our Amazingly Resilient Planet
67	Our Multi-Directional Genetic Code
292	Parrot to Testify at Murder Trial?
159	Peer-Review Fraud Strikes Again and Again
47	Penguins with Sunglasses
37	Planets Vanish!
161	Planned Parenthood's Roots in Darwinism
8	Plants Really Do Talk to Each Other
104	Poetic Verses on Multiverses
72	Prima Ballerina of the Sea
205	"Primitive" Electric Eel Not So Primitive After All
123	Problem Child
39	Purple Thief, The
248	Real Goal of Evolution, The
101	Satellite Image of the Exodus?
173	Science Makes the Case for God
78	Science of Parking Your Car, The
252	Science Turns to God's Design for Data Storage
154	Science Was Wrong About Boa Constrictors
185	Scientific Dissent from Darwinism, A
40	Scientist Fired for Dinosaur Discovery
26	Scientists Build the Universe
146	Scientists Still Don't Have a Clue How Life Began
203	Scientists Warn of Impending Crisis
294	Screws and Nuts in a Weevil's Legs
291	Second Skin for Soldiers
25	See Dick and Jane Evolve
52	Seven Flagellar Motors in One
22	Shooting at Evolution's Clay Pigeon
94	Silence of the Owls, The
81	Six Hundred Years Before Darwin
130	Sixty-Seventh Book of the bible, The
250	Slightly Imperfect Sale!
98	Smallest Life Form Discovered
178	Smithsonian Scientist Scolds National Geographic Society
151	Snake with Four Legs?, A
215	So Where Is Everybody?
131	Solving the Distant-Starlight Dilemma
274	Something's Fishy About Fish Evolution
89	South-Pointing Carriage, The
153	Spider Performs Amazing Engineering Feat
202	Spider That Thinks It's a Scuba Diver, The
14	Strongest Material in the World, The

83	Stupidity Virus, The
88	Tall, Fat and Upside-Dawn
113	Tech Titans Defy Death
45	"That's Evolution for You!"
212	"Then I Got Invited to a Bible Study"
240	Theory, Hypothesis … Or Something Else?
87	There's Something Fishy About This Car
149	This Plant Calls Out to Bats!
201	Those Strange Nazca Lines
266	Tiny Flies Inspire Big Advances in Hearing Aids
44	Tisha B'Av
258	Topsy-Turvy, Lefty-Righty
62	Trash-Talking Bats
122	Umbrella Bird, The
253	Unbelievable Migrations
29	Universe Not Expanding After All?
105	Unsolved Origin of Species, The
33	Vanilla Stumps Evolutionists
192	Virtual Reality – Blessing or Curse?
80	War Between Science and Faith?
281	Warning: Space Travel May Be Hazardous to Your Health
58	"Was I Going to Be Arrested?"
116	Water, Water Everywhere!
288	Waters Shall Be Turned to Blood
243	What Did the First Living Cell Eat?
132	What Has Evolution Given to the World?
61	What If Noah's Ark Is Found?
236	What Really Wiped Out the Dinosaurs?
200	What Your Tears Reveal About You
124	What's the Right Answer?
144	What's This? A Warm-Blooded Fish?
286	What's the Smallest Thing You Can See?
7	When Being "Wrong" Is Right
28	Where Did All the Water Go?
174	Where Did the Elephant Get Its Trunk?
82	Where the Battle Rages
257	Which Came First – The Orchid or the Moth?
262	Who Invented the Decimal Point?
264	Who Invented the Periodic Table?
211	Who's Looking Out for You?
254	Why Did God Give Us Fingernails?
93	Why Did God Give Us Fingerprints?
237	Why Do We Sigh?
295	Why Do We Smile?
135	Why Eve Was Made from Adam's Rib
77	Why You Were Born and Not Hatched
136	Will It Blend?
228	Windows of Heaven

38	World's Best Smelling Animal
35	World's Ugliest Animal?, The
170	World's Top Intellectuals Are Theists
220	You Are Mostly Bacteria
92	You Don't Use Science to Test Scripture!
209	Young Blood
48	Your Potato Chip Bag is Listening

About the *Letting God Create Your Day* Series

This volume of *Letting God Create Your Day* is one of several books published by Creation Moments supporting what the Bible teaches about creation. For more information about the other volumes in this series, call 1-800-422-4253 during regular office hours (Monday - Thursday, Central Time) or visit us online at CreationMoments.com.

Quantity discounts are available to churches wishing to give any of our *Letting God Create Your Day* books to students who attend their church. Ask about this discount when you call.

Contact us to request:

- Biblical creation DVDs and books, including other volumes in the *Letting God Create Your Day* series.

- "Creation Moments" CDs – each containing 30 radio programs.

- Information about the many free resources available from Creation Moments.

> **Creation Moments**
> **P.O. Box 839**
> **Foley, MN 56329**
> **1-800-422-4253**
> **www.creationmoments.com**